The Converging Technology Revolution and Human Capital

SOUTH ASIA DEVELOPMENT FORUM

The Converging Technology Revolution and Human Capital

Potential and Implications for South Asia

SAJITHA BASHIR, CARL J. DAHLMAN,
NAOTO KANEHIRA, AND KLAUS TILMES

 WORLD BANK GROUP

South Asia Development Forum

Home to a fifth of humankind, and to almost half of the people living in poverty, South Asia is also a region of marked contrasts: from conflict-affected areas to vibrant democracies, from demographic bulges to aging societies, from energy crises to global companies. This series explores the challenges faced by a region whose fate is critical to the success of global development in the early twenty-first century, and that can also make a difference for global peace. The volumes in it organize in an accessible way findings from recent research and lessons of experience, across a range of development topics. The series is intended to present new ideas and to stimulate debate among practitioners, researchers, and all those interested in public policies. In doing so, it exposes the options faced by decision-makers in the region and highlights the enormous potential of this fast-changing part of the world.

Contents

Boxes

Figures

Tables

Foreword

Human capital is South Asia's most important asset. How well this human capital is nurtured, utilized, empowered, and invested in will determine how strong a leadership role South Asia will play in the global economy during the twenty-first century. Over the last two decades, there has been reason for optimism. The region posted the fastest decline in under-five mortality and the fastest increase in life expectancy, made strong gains toward universal primary enrollment, and saw a rapid reduction in the proportion of the population living below the poverty line. But the shocks imparted by the first and second waves of the COVID-19 pandemic revealed the fragility of these gains, many important gaps in human development service delivery, and the vulnerability of large sections of the population stemming from deep structural inequalities. The shocks also highlighted the need for greater resilience as the region faces possible new waves of the pandemic and increasingly severe shocks caused by climate change and environmental degradation. This report thus comes at an opportune time as the region looks to rapidly rebuild its human capital.

The most important message of this report is that the ongoing converging technology revolution being spearheaded by the private sector in South Asia and globally should be exploited deliberately and strategically through public policy to build and protect, deploy and utilize, and empower human capital. The convergence of different groups of technologies, underpinned by universal connectivity, big data, and high-speed computing power—and propelled by artificial intelligence—is unleashing technological advances at an unprecedented speed. Specifically, these advances in technology can be used to improve access to and the quality of service delivery, prepare the workforce for technological changes in the workplace, encourage innovation for human development, and empower marginalized segments of the population.

The report lays out a framework for understanding the relationships between technology and human capital in service delivery, employment and innovation, and inclusion and empowerment. Several methods are used to assess the potential and implications of specific characteristics of the converging technology revolution for the region. Among other things, the report undertakes a landscape review of the technologies for service delivery in South Asia, highlighting the opportunities for rapid increase in coverage and quality through differentiation and personalization of services and the risks of deepening inequality due to lack of digital access. Importantly, it also reviews the implications of the converging technology revolution in the utilization of human capital in employment because automation and data-driven technologies are poised to cause extreme disruptions in a range of industries and services.

Meanwhile, South Asia's working-age share of the population is rising rapidly and will peak in 2040. The key challenge in realizing the demographic dividend is to make productive use of this population in the face of rapid technological change. This report highlights the opportunities for preparing and training the workforce, leveraging technology for improving the efficiency and effectiveness of service delivery in health, education, and social protection, and deliberately using the innovation ecosystem to create technologies for the human development sectors.

Empowering human capital is a major theme of this report, which sharply delineates the risks that the converging technology revolution, if not managed well, pose to the marginalized and vulnerable populations in South Asia. These risks arise from the undermining of trust and the lack of adequate data protection safeguards and weak structures for technology governance.

The rapid proliferation of technologies presents policy makers with many challenges. The temptation is to "pick" technologies that seem promising. And yet this study, drawing on discussions with technology experts, scenario exercises, analysis of the World Bank's human development portfolio, as well as the landscape reviews, eschews recommendations on specific technologies. Instead, it proposes action areas that are embedded in the defining characteristics of the converging technology revolution.

This report contributes a wealth of ideas and improves our understanding of how the ongoing technological revolution can be leveraged for accelerating human capital outcomes. It deserves the attention of policy makers, development practitioners, and others seeking to contribute to the South Asian century.

Lynne Sherburne-Benz
Regional Director for Human Development
South Asia Region
The World Bank

Acknowledgments

This report was produced by a World Bank study team led by Sajitha Bashir, adviser, Office of the Global Director for Education, under the management of Lynne Sherburne-Benz, regional director for human development, South Asia Region. The report was coauthored by Sajitha Bashir, Carl J. Dahlman, Naoto Kanehira, and Klaus Tilmes.

The full study team comprised as well the following people: Prasanna Lal Das, Rachel Halsema, Yusaku Kawashima, Stela Mocan, Eleonore Pauwels, Prema Shrikrishna, and Keong Min Yoon. Other members were Shuyang Huang, Luke Miller, Martin Moreno, Koshi Murakoshi, Joanna Sundharam, and Long Wang, who provided excellent research support. Aki Enkenberg contributed to many team discussions during the study. Sandra Alborta, senior program assistant, provided efficient administrative support to the team throughout the study.

The study team appreciates the guidance on the concept note and the feedback on initial presentations provided by the South Asia Regional Human Development management team: Lynne Sherburne-Benz, who provided leadership and encouragement throughout the study; Cristian Aedo, practice manager, Education; Trina Haque, practice manager, Health, Nutrition, and Population; Stefano Paternostro, practice manager, Social Protection; and Gail Richardson, practice manager, Health, Nutrition, and Population.

Special thanks to Hana Brixi, director, Human Capital Project (HCP), and her team—Zelalem Yilma Debebe, Ramesh Govindaraj, Amer Hasan, and Kelly Suzanne Johnson—for their insights on the HCP framework and earlier work on human capital and technology. This report was reviewed by Hana Brixi and William Colglazier, senior scholar, Center for Science Diplomacy, American Association for the Advancement of Science (AAAS), and editor-in-chief, *Science & Diplomacy*.

The team gratefully acknowledges the contributions of the over 100 external and internal colleagues who were consulted during the study. The team conducted a series of virtual meetings and workshops with policy makers, academics, research institutions, private sector entrepreneurs, and civil society organizations. They included the following: for Bangladesh—Javed Khan, Abdul Mannan, and Tom Crawford

(Virginia Tech); from Kerala, India—V. K. Damodaran (chairman, Foundation for Advanced Clean Energy Studies; chairman, Centre for Environment and Development; founder and director, Energy Management Centre; director of science and technology and environment, government of Kerala; leader, Kerala Sastra Sahitya Parishad/Kerala People's Science Movement); C. Balagopal (founder, Terumo Penpol); N. T. Nair (chief editor, *Executive Knowledge Lines*; founder, CMS Computers); John Samuel (president, Institute of Sustainable Development and Governance; civic leader); Amarnath Raja (executive chairman, InApp Technologies, Technopark); George Mathew (chairman and managing director, Team Sustain Ltd. and Solar + Clean Energy Technologies); Mini Thomas (director, National Institute of Technology, Trichy); and V. J. Kurian (managing director, Cochin International Airport Limited); from Nepal—Swarnim Waglé (chairman, Institute for Integrated Development Studies [IIDS]); Suresh Manandhar (artificial intelligence scientist, machine learning and natural language processing); Som Paneru (social entrepreneur; president, Nepal Youth Foundation and Ullens Education Foundation); Anand Bagaria (founder, NIMBUS, animal feed manufacturing company); Prativa Pandey (founder and CEO, Catalyst Technology and Herveda Botanicals); Himal Karmacharya (founder and president, Leapfrog Technology); Arnico Panday (CEO, Ullens Education Foundation); Bal Joshi (fintech entrepreneur); Arun R. Joshi (founder and CEO, Leadership Academy; head, Center for Human Assets, IIDS); Sumana Shrestha (head of global operations and strategy, Fusemachines); Lhamo Yanchen Sherpa (medical doctor, epidemiology); Jayendra Rimal (COO, Leadership Academy); Bishal Dhakal (founder, "Health at Home" delivery platform); Sandeep Sharma (principal and director, Adhyayan School); Parashu Nepal (founder and managing director, International Development Institute); Suman Timsina (executive director, International Development Institute); Pooja Pandey Rana (director of programs, Helen Keller International); and Dipta Shah (CEO, 54i Ventures); from Pakistan—Asif Shahid Khan, Athar Osama, and Mosharraf Zaidi; and Sriram Bharatam (founder and chief mentor, KUZA One), Kenya.

The study team benefited from inputs and advice from colleagues across the World Bank Group. These included Mehrin Ahmed Mahbub, Tenzin Lhaden, Hideki Mori, Jasmine Rajbhandary, and Deepika Chaudhery, Country Management Units, South Asia Region; Camilla Holmemo, Hadia Samaha, Keiko Inoue, Rene Solano, and Lire Ersado, program leaders for human development, South Asia Region; Robert Hawkins, Mike Trucano, and Karthika Radhakrishnan, Education Global Practice; John Blomquist, Ambrish Shahi, Shrayana Bhattacharya, Kenichi Nishikawa Chavez, Phillippe George Leite, Thomas Walker, Endeshaw Tadesse, Anastasiya Denisova, and Qaiser Khan, Social Protection and Jobs Global Practice; Paolo Belli, Meera Shekar, Kyoko Okamura, Kweku Akuoko, Lisa Saldanha, Rahul Pandey, and Dominic Haazen, Health, Nutrition, and Population Global Practice; Harsh Kapoor, Environment Global Practice; Taku Kamata, Alex McPhail, and Kris Welsien, Water and Sanitation Global Practice; and Parmesh Shah, Agriculture Global Practice. The team also received feedback from the human development teams working in Bangladesh, Bhutan, and Nepal on the findings of the draft report.

A wide range of participants from across the World Bank (WB) and external organizations took part in the six virtual scenario exercises. They included William Colglazier (American Association for the Advancement of Science); Rahul Chandran (Care International); Giulio Quaggiotto (head, Strategic Innovation, United Nations Development Programme); Rohinton Medhora (CEO, Center for International Governance Innovation); Yuka Shikina (external, education expert); Lynne Sherburne-Benz (WB, director, Human Development, South Asia Region [SAR]); Cecile Fruman (WB, director, Regional Integration, SAR); Hans Timmer (WB, chief economist, SAR); Nicole Klingen (WB, practice manager, Digital Development); Goran Vranic (WB, Global Regulation, Equitable Growth, Finance, and Institutions); Shobhana Sosale (WB, Education, SAR); Stela Mocan (WB, manager, Technology and Innovation Lab); Abhas Jha (WB, global lead, Disaster Risk Management); Juan Baron (WB, Education, SAR); Kenichi Nishikawa (WB, Social Protection, SAR); Ramesh Govindaraj (WB, Human Capital Project, SAR); Vyjayanti Desai (WB, practice manager, Identification for Development/Government to Person [ID4D/G2P]); Anastasia Nedayvoda (International Finance Corporation [IFC], Telecoms, Media, and Technology); Victor Mulas (WB, Innovation, Tokyo); Jonathan Marskell (WB, ID4D); Rahul Pandey (WB, Digital Health, SAR); David Wilson (WB, director, Health, Nutrition and Population); Aminata Niaye (WB, Digital Development, SAR); Cristian Aedo (WB, practice manager, Education, SAR); Siddhartha Raja (WB, Digital Development, SAR); Aaron Buchsbaum (WB, ID4D); Tina George (WB, Social Protection, SAR and Africa Region); as well as the authors of this report. The scenario sessions were facilitated by Eleonore Pauwels and Klaus Tilmes.

Finally, Crisy Meschieri contributed the design of several diagrams as well as the front cover design.

About the Authors

Sajitha Bashir is adviser in the Office of the Global Director of the Education Practice in the World Bank. Formerly, she was education practice manager responsible for over 20 countries in Africa. She currently leads the work on digital skills as part of the Digital Economy initiatives of the World Bank, where she has worked extensively on developing science and technology initiatives. She pioneered the World Bank's support for science and technology in Africa, including financing the Africa Centres of Excellence regional project and establishing the first pan-African science fund, the PASET Regional Scholarship and Innovation Fund. She is the coauthor of *Facing Forward: Schooling for Learning in Africa*, a flagship study of the World Bank on the quality of basic education in Africa. She has worked on projects related to Africa, Latin America, South Asia, and the Middle East. She holds a PhD in economics from the London School of Economics.

Carl J. Dahlman is currently senior policy adviser at the Growth Dialogue. He was chief economist and head of the Thematic Division of the Development Centre at the Organisation for Economic Co-operation and Development (OECD) between 2013 and 2016. At OECD, he was in charge of the biannual publication *Perspectives on Global Development*. He joined OECD in September 2013 after serving as associate professor at Georgetown University's School of Foreign Service from 2005 to 2013. Before that, he spent 25 years at the World Bank in various research, policy, and managerial positions, including resident representative in Mexico (1994–97); staff director, *World Development Report 1998/1999: Knowledge for Development*; manager and senior adviser, World Bank Institute (1999–2004); division chief, Private Sector Development; and division chief, Industrial Strategy. He has an extensive publishing record, including 12 books and numerous articles. He holds a BA in international affairs from Princeton University and a PhD in economics from Yale University.

Naoto Kanehira is senior strategy and operations officer at the Office of the World Bank Group Human Resources Vice President, coordinating the institution's senior management initiatives. He has led and contributed to operations and initiatives on science, technology, and innovation (STI) at the corporate, country, regional, and global levels at the World Bank, co-led the work on STI for the Sustainable Development Goals (SDGs) Roadmaps at the UN Inter-Agency Task Team on Science, Technology and Innovation, and served as the secretariat for the Bank's COVID-19 vaccine delivery taskforce. He is also a visiting scholar at the Center for Science Diplomacy, American Association for the Advancement of Science (AAAS), and a steering group member for the OECD on STI financing for the SDGs. Prior to joining the World Bank, he cofounded a mobile internet software start-up in 1998 and worked for McKinsey & Company from 2000 to 2010. He holds an MPA from Harvard University and an MSc in management from the Massachusetts Institute of Technology.

Klaus Tilmes is a senior policy adviser with more than 30 years of experience in development. He served as director of the Trade and Competitiveness Global Practice at the World Bank, overseeing regional operations in Africa and the Middle East and global teams for Trade and Competition Policy as well as Innovation and Entrepreneurship. Working with the President's Office, he was instrumental in developing the institution's strategy on disruptive technologies. During his World Bank career, he held operational and corporate positions, including director of strategy and operations, Finance and Private Sector Development Network; Knowledge Strategy adviser; manager, Independent Evaluation Group; and senior country economist. He also advises on the United Nations' Science, Technology, and Innovation program and government agencies in Africa and across the globe. In addition, he serves on the Sustainability Roundtable of the National Academies of Sciences, Engineering, and Medicine and as senior fellow of the African Center for Economic Transformation.

Executive Summary

South Asia's human capital challenges are among the most serious in the world. They include high levels of child malnutrition, deep deficits in early learning, low educational attainment, low life expectancy, an ongoing infectious disease burden, the disempowerment of women, and pervasive structural inequalities. Since the onset of the COVID-19 pandemic, recent gains in human capital outcomes have been reversed, with catastrophic consequences for the most vulnerable in the population. After the first wave of the pandemic, learning-adjusted years of schooling were projected to decline by 0.5 years, millions of jobs were lost, and infant and maternal mortality were expected to rise because of the disruption in health services. Meanwhile, environmental degradation and climate change pose devastating risks through their pervasive effects on health and the likelihood of mass displacement.

This study examines the potential contributions of converging technologies to the acceleration of human capital outcomes, their use and adoption during the COVID-19 pandemic, the risks associated with them, and the possible next steps for consideration by the World Bank.[1] As its starting point, the study used an enhanced framework of the World Bank's Human Capital Project and the analysis of the South Asia Human Capital Plan developed by the World Bank's South Asia Region unit. The plan identifies the three drivers that limit human capital outcomes: the poor quality and effectiveness of services, inequality of opportunities, and the increasing vulnerability of South Asian countries to a spectrum of shocks and risks.

The study team relied on four means of understanding the relationship between human capital and technology in general and converging technologies in particular: (1) analyses of the technology landscape in South Asia for service delivery in the human development sectors, the deployment of converging technologies in employment and innovation, and the safeguards and protections for the use of converging technologies in relation to human capital development; (2) interviews with external experts and World Bank staff; (3) a systematic examination of the technology components of the World

Bank's human development portfolio in South Asia; and (4) a scenario exercise involving external and internal experts and development practitioners to explore alternative futures for human capital outcomes in the region, taking into account broad technology metatrends and the critical uncertainties expected to influence outcomes. This overall analysis sets the stage for a series of recommendations for governments, development partners, and other stakeholders on using technology for accelerating human capital, as well as follow-up actions within the World Bank.

The Converging Technology Revolution and Human Capital

The term *converging technologies* refers to the synergistic combination of four groups of technologies: information technology, biotechnology, nanotechnology, and cognitive technologies. They go beyond digital technologies, although they are underpinned by the latter. The term was selected by the study team to move away from a siloed approach to individual technologies because it is precisely their combination that distinguishes the current technology revolution from preceding ones.

Data are central to the converging technology revolution. The integration of data from the human, physical, biological, and cyber spheres, in combination with high-speed computing power and connectivity and artificial intelligence (AI), are the other factors powering this revolution.

Although designed for beneficial purposes, converging technologies exhibit functions that can be easily misused and thus have a dual-use potential through interconnected systems, data flows, and the use of artificial intelligence. The most significant aspect of this revolution is the ability to affect the essence of human identity through human–machine augmentation and enhanced cognitive capacity. These trends raise complex governance issues related to agency, trust, and accountability, as well as moral and ethical issues.

The study team has articulated three dimensions of human capital and its interactions with technologies: (1) *building and protecting human capital* in all stages of the life cycle through health, education, and skills, as well as income; (2) *deploying and utilizing human capital* in the labor market and society at large; and (3) *empowering human capital.*

Human capital and technology interact in multifaceted ways. First, technology can accelerate the buildup and protection of human capital through health, education, and social protection services. The converging technology revolution is an opportunity to reimagine the delivery of these services, aiming for greater effectiveness (targeting, customization, and personalization) and efficiency. Second, technology can augment human capital outcomes through improvements in other sectors, such as food and nutrition, clean water and sanitation, electricity, transportation, digital infrastructure, and technologies that help improve the environment or mitigate climate change risks. Third, automation and data-driven technologies are changing the nature of human–machine interactions. Technologies used in manufacturing and services affect the deployment and utilization of human capital by altering the demand for the education and skills needed to absorb and deploy new technologies and reskill workers made redundant. Fourth, the convergence paradigm is changing the nature of science and innovation through the use of big data, AI, interconnected networks, and high-speed computing. Making use of this potential for new discoveries requires specialized human capital (such as technologists and scientists) for the creation and adaptation of technologies.

This framework also helps to highlight concerns about inequality and empowerment, which are central to the analysis of human capital. The diffusion and adoption of new technologies tend to favor, particularly in the first round, the more educated, who enjoy greater access to financial and other complementary assets, thereby increasing inequalities. The dual-use nature of these technologies also means that data collected for beneficial uses, such as personalized learning or medical diagnosis, can also be used for predictive behavioral surveillance, data manipulation, and targeted misinformation by public and private actors using these data. Furthermore, the ability to augment human capabilities using embedded technologies or enhanced cognitive capacities can help to reduce or widen inequality in human capital outcomes and power relationships, depending on who deploys these technologies and for what purposes.

BUILDING AND PROTECTING HUMAN CAPITAL: THE TECHNOLOGY LANDSCAPE FOR SERVICE DELIVERY IN SOUTH ASIA

The adoption of converging technologies in the delivery of services *to build and protect human capital* is already apparent in South Asia. The technology landscaping analysis reveals that the use of digital technologies in health and education service delivery is advancing in many South Asian countries, with India in the lead, especially in the private sector. The promise of converging technologies is greatest in the health sector, drawing on global technology trends and the greater standardization of diagnostic and treatment procedures across countries. In education, the focus is much more on digital learning tools, which are circumscribed by adherence to local curricula and examinations. In health and education technology, the private sector is taking the lead, with private health providers and educational institutions serving wealthier households. India's tech start-ups in education are becoming significant even globally. Even so, technology applications are still focusing largely on tutoring and examination preparation for students in the formal school system rather than on revolutionizing the approach to learning, reaching the out-of-school population, or upskilling informal sector workers.

Digital public platforms have been developed in some South Asian countries for health, education, and social protection, but they are lagging in their effective use by beneficiaries and are not yet transforming service delivery. Among the barriers to their effective use is the limited "first-mile" access to digital infrastructure (reliable, high-speed, and affordable connectivity and devices) for schools and health centers in poor communities, as well as for the women in households. Less than 40 percent of women own a mobile phone in India and Pakistan, compared with 80 percent of men in India and 70 percent in Pakistan. Another barrier is the limited availability of local content, which is especially important in education, but also for delivery of health services. Unless the low skill levels of beneficiaries and the barriers impinging on the uptake of digital technologies by end users—such as teachers, students, and frontline health workers—in poor, marginalized communities (or those excluded from the traditional power structures, especially women and girls) are taken into account, there will be no significant change in the quality of the services or the experience for the ultimate beneficiaries.

The first wave of the COVID-19 pandemic saw a rapid shift toward digital delivery of human development services, with notable advances in accelerating digital payments for scaled-up social assistance programs. However, it also highlighted the deepening of pre-existing inequalities in human capital. Although the private health and education sectors were able to pivot to digital service delivery, public health and education services struggled to provide services to the poor, with limited platform readiness even when available. A stark example is in education, where most children in public schools had limited or no access to education for most of 2020, resulting in high learning losses and dropout rates, especially among girls. Meanwhile, those in private schools and universities benefited from a shift to online and mobile learning. Data-driven decision mechanisms, using geospatial technologies and the Internet of Things, are relatively limited in the public sector, although they could play a critical role in improving the resilience of service delivery.

The use of more sophisticated converging technologies, underpinned by AI-enabled analysis of large datasets, is occurring without adequate safeguards in South Asia, especially in health and education. There is a risk of large-scale harm to populations through possible targeting and deliberate exclusion of specific groups, exfiltration of data for other purposes, or cybersecurity vulnerabilities. The use of these technologies among children and populations with limited literacy or technology awareness should be regulated to avoid deepening of inequality, disempowerment, and the growing trust deficit that could undermine social cohesion.

DEPLOYING AND UTILIZING HUMAN CAPITAL: IMPLICATIONS OF THE CONVERGING TECHNOLOGY REVOLUTION FOR EMPLOYMENT AND INNOVATION

Converging technologies will have a profound impact on the second dimension of human capital, *the deployment and utilization of human capital* both at work and in the innovation system in South Asia. Regional estimates for job losses and new job creation stemming from technology adoption are relatively scarce and highly variable across sectors. Studies suggest that the impact of automation and robotics could be greatest in the capital-intensive sectors, which account for a relatively small share of jobs. Labor-intensive industries (such as textile, leather, and footwear) are unlikely to be rapid or large-scale adopters of automation technologies because of the high capital investment required as well as the current labor cost advantages. However, there is a risk that major brands in the ready-made garment industry may relocate their manufacturing hubs closer to their markets. Agriculture has a high potential for automation through the use of sensors, digital platforms, and data analytics, but the declining size of landholdings and the limited skills of farmers make widespread technology adoption unlikely. The informal service sector similarly faces constraints in the adoption of technologies, but new jobs could be created through e-commerce platforms. Although the scale of the impact on employment is difficult to gauge, job losses and displacement are likely to be significant, raising the need for massive reskilling and adaptive social protection, including for informal sector workers.

Innovation is being transformed by the penetration of converging technologies, in particular the reliance on big data, massive data storage capacity, high-speed computing power, and AI to generate scientific discoveries and translate them into technological applications. However, apart from India, innovation ecosystems in other South Asian countries are nascent. Although India has the fourth-largest number of technology "unicorns" in the world, it lags far behind the United States and China. Among other things, countries in the region lack the advanced, specialized human capital to adapt existing technologies to national and local needs, as well as the firms that can use them. In terms of research and development expenditures, India accounts for just 4 percent of the global total (compared with China, which accounts for over 23 percent), while Pakistan and Bangladesh account for 0.3 percent and 0.2 percent, respectively. South Asian countries also lag in the accumulation of advanced human capital as shown by the slow growth in the tertiary education ratio, advanced degrees in natural sciences and engineering, and number of researchers. Researchers lack access to large-capacity networks, high-performance computers, advanced digital skills, and means of collaboration to take full advantage of the digital revolution in science and innovation. Higher education curricula and teaching methods in the region have not fully embraced the requirements of the convergence revolution in terms of cross-domain programs in universities—that is, passing on to students the knowledge, tools, and ways of thinking that they need from the life and health sciences; the social sciences; the humanities and the arts; the physical, mathematical, and computational sciences; and the engineering disciplines to address scientific and societal challenges.

At the same time, digital technologies are also enabling the democratization of innovation with the spread of community "fablabs" and local networks of professionals and entrepreneurs linked to global networks to share knowledge and practices. Examples of these local innovation hubs abound in South Asia, but they are still relatively small in scale and suffer from lack of funding and technical support.

HUMAN CAPITAL EMPOWERMENT: THE IMPORTANCE OF TRUST, DATA SAFEGUARDS, AND PROTECTION OF VULNERABLE GROUPS

The third dimension of human capital, *empowerment*, is especially vulnerable to the converging technology revolution because of its dual-use characteristics. Technologies operate within a sociotechnical system of values, norms, cultures, laws, and regulations. Trust underpins the deployment and use of technology and is built on the perception that technology delivers tangible benefits to people. In South Asia, where harassment, violence, and disinformation directed at minorities and women are rampant, many studies have documented a worsening of these trends through social media. Trust is undermined through massive disinformation that may incite violence and invasive private and public data protection systems. This erosion of trust is exacerbated by the exclusion of "data invisible" groups (such as minorities or tribal communities) and the digital abuse and exploitation of women and children.

The systems of regulation and governance to deal with converging technologies and the paradigm shift in data and its uses have yet to be put in place globally. In South Asia, the data and technology governance framework is still nascent or almost absent in some countries. In the context of deep-rooted structural inequalities, these frameworks urgently need to be strengthened. Although there has been some progress in developing legislative bills and acts in this area, critical trust-building provisions remain unaddressed, including on data proportionality, misuse, and misappropriation. Meanwhile, institutional mechanisms for enforcement, accountability, and grievance redressal are lacking, and many of the laws contain exceptions for government surveillance and emergencies, with a low burden-of-proof threshold.

Technology in the World Bank's Portfolio of Human Capital Projects

The World Bank portfolio of ongoing and pipeline human capital projects in South Asia included as of June 2020 a significant allocation for technology components (US$6.4 billion out of US$15 billion in project commitments). An overwhelming share of these technology components is focused on the build and protect dimension of the human capital framework used in this study, through improved service delivery. A relatively small proportion pertains to the deploy and utilize dimension to prepare for technological changes in the workplace and to adapt and develop new technologies. The investments related to the empower pillar are even less and of relatively recent origin to safeguard against technology risks and proactively protect the vulnerable in the population. About 60 percent of technology investments are at the piloting stage, with 30 percent for scaling technology adoption and only 10 percent for systemic transformation. Projects in some countries and in sectors with more advanced digital capabilities (especially social protection) have implemented targeted interventions for scaled and systemic technology adoption such as through government platforms and public-private ecosystems. Technology interventions in the project portfolio do not include many converging technologies that are already part of the technology landscape for service delivery in South Asia.

Scenario Planning: Imagining Alternative Futures for Human Development in South Asia

The scenario exercise, undertaken with a broad array of internal and external experts in mid-2020, explored alternative futures for human capital outcomes, depending on how different economic, social, and political factors interact with technology trends. The exercise considered the interaction of four technology metatrends that characterize the convergence revolution and two critical uncertainties expected to influence the near-term social and political context in the region. The four technology metatrends are (1) technologies for building and protecting human capital by improving service delivery in health, education, and social protection, as well as in contributing sectors; (2) the impact of data-driven and human–machine production technologies on the demand for and use of human capital and on economic and social structures; (3) the increasing importance of dynamic innovation ecosystems for human development to producing, adapting, and diffusing technologies to address local needs; and (4) the need to develop governance arrangements for converging technologies to exploit the benefits and mitigate the risks they create for human development. At the time this scenario was developed in mid-2020, uncertainties were dominated by the interplay of two factors: the trajectory of the pandemic and severity of its impact on human capital outcomes, on the one hand, and the countervailing or mitigating dynamics of cooperation versus isolation in responding to the unfolding crisis at the global, regional, national, and local levels, on the other, which may affect domestic social cohesion and regional and international cooperation.

The most optimistic scenario portrayed a future with inclusive technology for human capital that improves service delivery for the poor, creates new jobs and local innovation, and empowers human capital. The other three scenarios offer a range of outcomes with varying degrees of challenges and, at the outer spectrum, outright pessimism. The latter reflects the devastating impacts of the pandemic and rising geopolitical tensions, with a breakdown in international solidarity and internal social cohesion. The key recommendations of the scenario exercise helped to inform the future areas of engagement presented in the next section.

Discussions with regional experts highlighted the accelerated adoption of digital technologies by governments and citizens in the delivery of human development services after onset of the pandemic. However, the ability to use technologies to develop local solutions depended on the level of community preparedness and concern for disadvantaged social groups. Strengthening the foundations for resilience and adaptability is urgent given the enormous risks to human capital in South Asia posed by technology-induced disruptions in employment and climate change and environmental degradation. Addressing these risks calls for building long-term capacity for leadership and local solutions by means of increasing the availability of skills, providing infrastructure and networks of innovators, and creating social infrastructure based on trust in the use of technologies.

Nine Action Areas: Leveraging Converging Technologies for Improving Human Capital Outcomes

Informed by these findings, this study avoids recommending specific technologies because their maturity level, sectoral specificity, and readiness for adoption differ across countries and sectors. Moreover, there are large uncertainties about how the technology metatrends will play out in the region. Instead, the study proposes nine action areas in which governments, development partners, the private sector, and local communities could shape the future and accelerate human capital formation in the context of the global converging technology revolution, the urgent need for a rapid recovery from the pandemic, and greater resilience in the face of climate change and environmental degradation.

The nine action areas in which to strengthen all dimensions of human capital are depicted in figure ES.1. These actions fall into three broad functional categories (the three wedge-shaped segments of the quarter circle in figure ES.1): (1) improving service delivery, (2) building future resilience and adaptability, and (3) promoting inclusion. These action areas can also be understood by their potential impacts, ranging from essential and cross-cutting to transformational (the concentric segments in figure ES.1). The first set consists of certain essential and cross-cutting actions fundamental to enabling impact across all three functions. A second set of actions enables customization and integration to have impact at scale. The third set of actions is more difficult to achieve but can have profound transformational impacts on all aspects of service delivery, building resilience and promoting inclusion. The suggested action areas are not meant to be addressed in a mechanical or isolated manner. Many are interdependent, and some may need to be undertaken first for others to be more effective. A common theme is the need to build trust among citizens by demonstrating the ability of technologies to solve their problems and to build the skills and capabilities in the public sector to safely use these technologies for the benefit of citizens.

Action area 1: First-mile equitable digital access. To benefit from digital and converging technologies in service delivery, the most vulnerable and marginalized populations need digital access. Access includes not only affordable broadband connectivity but also availability of devices and foundational digital skills. Equitable access includes access for women and girls in households and for deprived communities. Proposed follow-up actions include creating digital road maps for first-mile digital access for all government departments, defining meaningful access standards (for broadband speed,

FIGURE ES.1 **Nine Action Areas in Which Technology Can Build and Protect, Deploy and Utilize, and Empower Human Capital**

Source: World Bank study team.
Note: AI = artificial intelligence; AR/VR = augmented reality/virtual reality; HD = human development.

devices, and affordability), devising programs for universal digital skills training, and using universal service funds to reach underserved communities.

Action area 2: Community participation and trust. Community participation in the deployment of technology can build trust and accountability in its use. Bottom-up innovation can produce locally adapted solutions, strengthen resilience in anticipation of crises, and create local jobs. The ready availability of technology, open-source platforms, and data-enabled communities also allows broad participation in innovation processes and the mobilization of local ecosystems of scientists and technologists, youth, local entrepreneurs, and government. Follow-up actions include learning from successful community participation movements and helping to create local ecosystem clusters that link regional and global networks, build the skills of local innovators, and strengthen trust and social cohesion through inclusive governance norms.

Action area 3: Public-private digital platforms for human development services. Public platforms (in partnership with the private sector) can serve as an "equalizer" to the inequality-expanding effects of many technology applications. The design and management of platforms require a user-centric, holistic approach that focuses on the needs, contexts, and constraints of the intended beneficiaries. Follow-up actions could include developing the public sector's capacity to create the appropriate policies and skills needed for developing and managing digital platforms for health, education, social protection, job matching, and reskilling.

Action area 4: Integrated, dynamic social registry. Integrated, dynamic social registries enable the adaptation and scaling up of social protection in times of crisis by increasing coverage or benefits, replacing cash with food where markets fail, setting aside conditionalities, and ensuring that vulnerable groups have ready access to benefits. Priority actions would include expanding the portability of benefits and income support for the large urban informal sector; coordinating and integrating the multitude of programs across government departments; building inclusive, citizen-friendly platforms; expanding use of digital payments; and integrating national identification systems with adequate data safeguards and protections.

Action area 5: Local inclusive digital content. South Asia is home to several hundred local languages, and yet Asian language content is almost completely absent online. This absence has implications for who benefits from technology and who and what get represented—and by whom. Access to local content offers the opportunity to bring and adapt knowledge and information to non-English speakers, to accelerate teaching of marginalized groups, and to create virtual linguistic communities. Follow-up actions would include building up local data repositories, creating local language access protocols, fostering a market for local digital content, and using local languages in consultations and knowledge sharing. Translation technologies can also open new markets and jobs.

Action area 6: Data-driven decision-making. Two data-intensive technologies can be transformational in targeting and optimizing service delivery and building resilience: geospatial technologies and the Internet of Things. Their effective use requires a high level of capability within government for decision-making. Possible interventions include the deployment of data-driven decision support mechanisms to identify and target services for population groups in locations facing malnutrition or illiteracy. A major thrust of follow-up actions would be building the capacity of key human development ministries in data analytics and data-based decision-making, as well as inducing a cultural change to focus on the experience of citizens in the design and implementation of programs.

Action area 7: Open Science. Open Science is an approach aimed at enabling all countries to benefit from the global digital revolution in science and the ongoing converging technology revolution, sharing knowledge across disciplines and societies. This approach complements and reinforces the building of local ecosystems for innovation (action area 2). The data revolution, combined with AI and high-speed computing power, can help address challenges that are inherently complex, including climate change and sustainable development. Specific action items include developing protocols and the appropriate governance structures for Open Science; strengthening national research and education networks; creating high-speed infrastructure, data storage and sharing mechanisms, and training in advanced scientific skills, including in data science; and incentivizing collaboration through funded collaborative programs on priority challenges.

Action area 8: Inclusive and open artificial intelligence. AI has great potential for personalizing learning and health services, job matching, optimizing systems, automating processes, and improving the efficiency of service delivery. However, the risks of disempowering human capital are also great: lack of agency in workplaces, worsening inequality and reinforcement of bias, exclusion, concentration of power, and surveillance. The lack of "explainability" of AI models undermines societal trust. Accelerating the development and adoption of AI for human capital empowerment and inclusion requires building a government's capacity to develop policies based on ethical principles, engaging with global AI networks to help develop standards and protocols, developing safeguards in the use of data, and enabling public scrutiny and peer review of AI models.

Action area 9: Technology and data governance. The safe, ethical use of converging technologies, including the data flows that underpin them, can be transformational in driving the other action areas, can promote trust, and can strengthen the empowerment of human capital. Human development sectors tend to lag behind other sectors in digital data protection, capture, and use, thereby limiting benefits for the majority of the population, while exposing them to the risks stemming from the unethical and criminal use of data. These risks are amplified because the South Asia region is one of the largest data markets globally, making it highly attractive to leading technology companies seeking to mine data to develop artificial intelligence and related technology applications. Development of the policy and regulatory framework for data across human development sectors should be a priority, balancing the goal of promoting beneficial use with that of creating adequate safeguards. Other actions include designating the datasets that should be made available as a public good, identifying the specialized data institutions and oversight mechanisms, analyzing technology and data value chains, and developing standards for accountability, transparency, and grievance redressal.

The World Bank also needs to build up its own capacity to take advantage of the potential of converging technologies for accelerating human capital outcomes, while ensuring inclusion and empowerment. Three broad sets of actions are therefore proposed for the World Bank: (1) develop partnerships to strengthen in-house knowledge in order to become a better-informed practice leader; (2) fast-track a shared understanding for technology-enabled human capital programs; and (3) develop human development's service offerings on technology design, advice, and delivery. The latter includes using frequent feedback from implementation, research, and interactions with stakeholders to improve and develop future action areas because converging technologies, their applications, their impacts, and their governance continue to evolve.

* * *

The deployment of converging technologies for human capital is set to soon expand. Because of the rapidly changing technology landscape, ongoing analysis and technology foresight, as well as scenario planning exercises with government agencies, the private sector, and community groups engaged in innovation, are needed to hear different voices and to sensitize all participants to what is at stake. The World Bank's policy dialogue and operational support for this agenda, embedded in its core mission and its ability to facilitate regional collaboration and consensus around common approaches and innovation, could help to shape the future in favor of strengthening human capital in all its dimensions and for the entire population.

Note

1. This report is largely based on analysis completed between March and October 2020. The study findings provided inputs for the preparation of the World Bank's South Asia Human Capital Plan. The time frame for this work means much of the data on the impacts of COVID-19 on human capital outcomes relates to the first nine months of the first wave of the pandemic.

Abbreviations

AI	artificial intelligence
BMGF	Bill and Melinda Gates Foundation
B2C	business-to-consumer
CSR	corporate social responsibility
DHIS2	District Health Information Software 2
DIKSHA	National Digital Infrastructure for Knowledge Sharing (India)
DPL	Development Policy Loan
GDP	gross domestic product
GNI	gross national income
GP	Global Practice
G2P	government-to-person
HCI	Human Capital Index
HCP	Human Capital Project
HD	human development
ICT	information and communications technology
ID	identification
IEEE	Institute of Electrical and Electronics Engineers
IoT	Internet of Things
MIS	management information system
MIT	Massachusetts Institute of Technology
NGO	nongovernmental organization

NREN	national research and education network
NSER	National Social and Economic Registry (Pakistan)
OECD	Organisation for Economic Co-operation and Development
P4R	Program-for-Results (financing)
R&D	research and development
RTBF	right to be forgotten
SAR	South Asia Region
SDG	Sustainable Development Goal
SPL	social protection and labor
STEM	science, technology, engineering, and mathematics
STI	science, technology, and innovation
UN	United Nations
UNICEF	United Nations Children's Fund
USAID	United States Agency for International Development
WFP	World Food Programme
WHO	World Health Organization

CHAPTER 1

Introduction

Introduction

South Asia's human capital challenges are among the most serious in the world. They include high levels of child malnutrition, deep deficits in early learning, an ongoing infectious disease burden, the disempowerment of women, and pervasive structural inequalities. With the onset of the COVID-19 pandemic, recent gains in human capital outcomes began to be reversed, with catastrophic consequences for the most vulnerable in the population. Environmental degradation and climate change pose even more devastating risks through their pervasive effects on health and their potential for mass displacement.

COVID-19 has significantly worsened the human capital outlook across South Asia.[1] Historical precedents, expert opinion, and the experience of countries in other regions affected by this pandemic suggest that further outbreaks of the disease are likely to continue in South Asia in new waves, with viral mutations from different parts of the world and the relatively slow global vaccine rollout adding to uncertainties about the pandemic's severity and duration.

By the end of August 2020, within a few months of the start of the first wave of the pandemic, the region had reported 4.4 million infections and 75,000 deaths (officially).[2] Highlighting the scale of direct and indirect impacts, preventive and curative health services were diverted to fight COVID-19, leading to poorer health outcomes for the most vulnerable. Almost 40 million children in Pakistan reportedly missed their polio vaccine drops after cancellation of the nationwide campaign in April 2020. Under-five mortality rates in all countries are likely to rise for the first time in decades. And across the region, deaths from tuberculosis, already one of the main causes of mortality, are

1

expected to rise because people are unable to be diagnosed or complete their treatment regimens.

In education, widespread school closures in 2020 led to a loss in learning, which was already at a low level prior to the pandemic. Lockdown measures in South Asia in the early months of the pandemic in 2020 were more stringent than those in Europe or North America, resulting in almost 400 million children out of primary and secondary schools, with a disproportionately negative impact on girls. An estimated 5.5 million students may drop out of the educational system. The projected learning loss for the region for the first five months of school closures after the onset of the pandemic, when remote learning opportunities were relatively limited, is 0.5 learning-adjusted years of schooling. Initial estimates put the loss in lifetime earnings per student at US$4,400, which is equivalent to 5 percent of total earnings (World Bank 2020).

Meanwhile, the pandemic has led to staggering job losses and disruptions for migrants, which will have longer-term economic consequences. About 50 million jobs have been lost in South Asia in the first phase of the pandemic.[3] The return to Nepal and Bangladesh of hundreds of thousands of migrants from India and to South Asia from the Gulf countries has had large negative multiplier effects on households, adding to the pool of job seekers. In a region with the lowest rate of social protection coverage in the world, these numbers have translated into an increase in poverty and insecurity for the most vulnerable.

The Priorities for South Asia

An immediate priority for the South Asia region is how to contain and recover rapidly from the shock to human capital outcomes and ensure rapid progress toward achieving better and more equitable outcomes. The response to the COVID-19 pandemic, globally and in the region, showed the important role that technologies are playing in different areas of the pandemic response—from the health sector response, including the accelerated development of new types of vaccines, to the use of digital technology in remote learning, to the rapid delivery of social assistance programs.[4] It also showed the differential impact of technologies on countries and population groups.

Moreover, the profound converging technology revolution that is engulfing the world, which predates the pandemic, is altering the relationship between technology and human capital, and making it more complex by creating vast new opportunities for accelerating human capital outcomes while generating new and considerable risks. This convergence revolution is characterized by the merging of virtual, physical, biological, and cognitive technologies with the power of big data, machine learning, and artificial intelligence (AI).[5] As a result, biomedicine and the health sector have seen major developments in imaging, biomaterials, nanotechnology, bioinformatics, cellular technology, and medical robotics, and AI is increasingly used in diverse areas. Digitization combined with AI is creating the possibility of mass personalized learning—anywhere, anytime. In agriculture, the tools of synthetic biology are being used to tailor food products to meet specialized dietary needs, and nanotechnology is being used to develop smart plant sensors that communicate with electronic devices to optimize the use of water, fertilizers, and pesticides. To protect the environment, scientists are developing biosensors to monitor environmental changes. Meanwhile, the use of new materials developed through nanotechnology, automation, robots, and intelligent manufacturing systems are revolutionizing the workplace and placing demands on the education and training sector for the creation of new knowledge and skills.

But beyond these positive developments, there are huge implications for human capital empowerment because data on individuals and social groups are being collected by multiple applications and devices, and AI is automating the analysis of data from different sources. A world of increasing automation in production and services, new discoveries through powerful machine-driven analysis and knowledge synthesis, and the delegation of decision processes to AI algorithms is well within reach over the coming years, and it is already beginning to shape today's reality in some more advanced situations. Thus the converging technology revolution holds the potential to fundamentally change the ways in which human beings live, behave, work, and interact.

This Study

The World Bank's Human Capital Plan for South Asia has served as an anchor for this study.[6] The plan identifies three key drivers that limit human capital outcomes in the region: (1) scarce public funding, which contributes to the poor quality and effectiveness of services; (2) multiple inequalities; and (3) increasing vulnerabilities to a spectrum of shocks and risks. Accordingly, the plan's strategic priorities are to "invest smarter and with quality; include and empower, especially adolescent girls and women; insure and prepare for potential shocks and risks; and innovate through data, technology and multi-sector action."

This study assesses the opportunities and implications of converging technologies for promoting human capital outcomes, including to reduce inequality and strengthen inclusion and empowerment, to identify the World Bank's current engagement in the use of technology in human development (HD), and to suggest priorities for future work.

This study uses a variety of methods, including a conceptual framing of the relationship between human capital and technology; desk-based reviews of the technology and policy landscape in South Asia; a systemic, technology-focused analysis of the World Bank's human development portfolio; scenario exercises; interviews with external experts; and discussions with World Bank staff.[7]

The report is structured as follows. Chapter 2 presents the features of the converging technology revolution, the theory of change linking technology with the human capital framework, and technology classification schema to analyze the technology landscape in South Asia and the portfolio of World Bank projects. The chapter concludes with views offered by external experts on future pathways for the use of technologies in the region. Chapter 3 reviews the technology landscape in South Asia, covering inequalities in digital access and digital infrastructure, the existing use of technologies for building and protecting human capital through human development services in the public and private sectors, and case studies illustrating the use of converging technologies in human development. Chapter 4 begins with an overview of the implications of these technologies for future jobs and skills and, in particular, for the innovation system and the requirements for advanced human capital. Chapter 5 then discusses technology and governance in relation to the empowerment of human capital, the role of trust, the implications for data policy and safeguards, especially for vulnerable populations, and accountability and citizen oversight over the deployment of AI and other dual-use technologies. An analysis of the technology components of the World Bank's HD projects in South Asia, covering the current active portfolio and pipeline of about US$15 billion, is presented in chapter 6. The conclusions and findings of the scenario exercises are covered in chapter 7, while the concluding chapter 8 presents the key recommendations.

Notes

1. This report is largely based on analysis completed between March and October 2020. The study findings provided inputs for the preparation of the South Asia Human Capital Plan. The time frame for this work means that much of the data on the impacts of COVID-19 on human capital outcomes relate to the first nine months of the first wave of the pandemic.

2. Data are as of August 31, 2020 (Our World in Data, https://ourworldindata.org/coronavirus -data). Because of limited testing and lack of health care facilities, the numbers are likely to be underestimated.

3. Estimated by the International Labour Organization as the difference between the pre-COVID-19 employment projection for 2020 and preliminary estimates for 2020 after the onset of COVID-19 (ILO 2020).

4. Because of the timing of the preparation and finalization of this report, the use of technology in the rollout of vaccine delivery in South Asia could not be covered here.

5. The concept of convergence is explored in several reports. See Bainbridge and Roco (2016); OECD (2020); and Roco et al. (2013).

6. See World Bank presentation, "Unleashing the South Asian Century through Human Capital for All: 4i4HCA: A Framework for Human Capital Acceleration in South Asia," http:// documents1.worldbank.org/curated/en/885781616013381140/pdf/Unleashing-the-South -Asian-Century-Through-Human-Capital-for-All.pdf.

7. Over 100 persons participated in the scenario exercises and expert discussions.

References

Bainbridge, W. S., and M. C. Roco. 2016. *Handbook of Science and Technology Convergence.* Berlin: Springer Reference.

ILO (International Labour Organization). 2020. *Asia-Pacific Economic and Social Outlook: Navigating the Crisis towards a Human-Centred Future of Work.* Geneva: ILO. https://www.ilo.org/wcmsp5/groups/public/---asia/---ro-bangkok/---sro-bangkok/documents/publication/wcms_764084.pdf.

OECD (Organisation for Economic Co-operation and Development). 2020. *The Digitalisation of Science, Technology and Innovation: Key Developments and Policies.* Paris: OECD Publishing.

Roco, M. C., W. S. Bainbridge, B. Tonn, and G. Whitesides, eds. 2013. *Converging Knowledge, Technology and Society: Beyond Convergence of Nano-Bio-Info-Cognitive Technologies.* Boston: Springer.

World Bank. 2020. *South Asia Economic Focus, Fall 2020: Beaten or Broken? Informality and COVID-19.* Washington, DC: World Bank. https://openknowledge.worldbank.org/handle/10986/34517.

CHAPTER 2

The Converging Technology Revolution and Human Capital

Introduction

This chapter sets out a framework for understanding converging technologies and the complex and mutually reinforcing relationships between human capital and technology. It begins by describing the distinguishing features of the converging technology revolution.[1] The current stage of technological change is often referred to as the Fourth Industrial Revolution (Schwab 2016). However, a more appropriate term is the *converging technology revolution* because the impact is not only on industry but also on services and because it involves intangibles comprising virtual models of the physical and biological world.[2]

The first section of this chapter explains how this study frames the multisided relationship between human capital and technology within the context of the World Bank's Human Capital Project (HCP).[3] This conceptual framework underpins the structure of the rest of this report. This chapter concludes by presenting the findings from interviews with technology experts and representatives of the private sector and civil society in the South Asia region as a prelude to the analysis presented in subsequent chapters.

Converging Technologies

The term *converging technologies*[4] refers to the synergistic combination of four groups of technologies: information technology, biotechnology, nanotechnology, and cognitive technologies. They go beyond digital technologies, although they are underpinned by the latter. The term has been selected by the study team to move away from a siloed approach to individual technologies because it is precisely their combination that distinguishes the current technological revolution from preceding ones.

Data from the human, physical, biological, and cyber spheres and their integration with these technologies are central to the converging technology revolution. High-speed computing power and connectivity are the other two factors powering this revolution. The development of artificial intelligence (AI) is further enabling and driving the converging technology revolution. AI is itself a combination of information technology and cognitive science, and it is now increasingly viable through the availability of vast amounts of data, cheap high-speed computing power, and ubiquitous connectivity.

The converging technology revolution has the potential to restructure the delivery of publicly and privately provided services for human development through personalization, precision targeting, cost reductions, and new organizational and accountability arrangements. Improvements in other economic sectors, such as agriculture, energy, water, and transportation, can have an indirect impact on improvements in human capital formation.

However, the risks are high. Converging technologies, because of their digital nature, operate in cyberspace, and thus cybersecurity is essential to protect populations and their personal data. Furthermore, although they are designed for beneficial purposes, converging technologies exhibit functions that can be easily misused and so could have a dual-use potential. Examples of dual-use applications include predictive behavioral surveillance, data manipulation, and targeted misinformation by the state and other actors using data collected in the context of otherwise beneficial uses, such as personalized learning or medical diagnosis.

The most significant aspect of this revolution for human capital is its ability to affect the essence of human identity through human-machine augmentation and enhanced cognitive capacity and thus to reduce or widen inequality in human capital outcomes and power relationships. Earlier technological revolutions also required an adjustment of political and economic structures, laws and regulations, public policy, and societal norms and cultures. With the accelerating speed of the converging technology revolution, there is strong evidence that these superstructures are unable to adapt quickly enough, with negative implications for the region's socioeconomic trajectories. In South Asia, where levels of human capital are already low, where public policy capacity is lagging, and where inequalities of gender, religion, caste, and community are deeply entrenched, the converging technology revolution could add to an already combustible social mix and lead to unpredictable consequences.

The converging technology revolution is transformational. Advances in science and better knowledge of the building blocks of matter and life are now permitting the creation of new materials with nanotechnology and altered or entirely new life forms with gene editing and gene drives. Data, an intangible, have become a critical new factor of production and value addition, while knowledge can be produced without direct human involvement and agency. In addition, the reach and dual-use characteristics of many of these technologies raise ethical, moral, social, and governance considerations that society has yet to address:

- **Ethical.** Who is responsible for deaths caused by autonomous vehicles? Is it ethical to produce clones of humans? Who is accountable for the negative impacts of new materials or new biological forms on the environment and people? How does society avoid discrimination arising from bias in algorithms or analysis of gene characteristics that may trigger adjustments in insurance rates based on predisposition to disease or block access to educational opportunities?

- **Moral.** What is life? How much should genes be manipulated to create new life forms, including babies with special characteristics?

- **Social.** How does society counterbalance the tendency toward rising inequality that appears to stem from the use of these technologies? Who has the right to personal data?

- **Governance.** What is the responsibility of the state in safeguarding citizens' data? How can surveillance conducted to maintain control by governments be minimized, and how do citizens have recourse to due process? What does "informed consent" mean when personal information is collected without a person's knowledge, and when few people understand the algorithms used for the predictive analytics that shape behavior?

Framing the Relationship between Human Capital and Technology

The World Bank's Human Capital Project has identified the key elements in building the human capital of the next generation, specifically through "allowing all children to reach their full potential—growing up well-nourished and ready to learn, attaining real learning in the classroom, and entering the job market as healthy, skilled, and productive adults" (World Bank 2016). The emphasis is on enabling households to develop their human capital through a supportive economic and political environment. As part of the HCP, the World Bank has developed the Human Capital Index (HCI), which measures the amount of human capital that a child born today can expect to attain by age 18.[5] The HCI currently excludes critical dimensions of human capital, especially of the current adult population and workforce, such as the skill levels of the workforce (which should ideally go beyond educational attainment to include job-relevant and socioemotional skills), the deployment and utilization of human capital in employment and society, and the health of the adult population, which can affect productivity and quality of life.

Various World Bank documents related to the HCP have articulated the dimensions of "building, protecting and deploying human capital" and have also highlighted the importance of empowerment (especially in relation to women), citizen engagement, and social accountability to enable households to build human capital (World Bank 2019). So far, the HCP does not incorporate technology as a catalyst or inhibitor for the development of human capital.

Building on these elements, the study team has articulated three dimensions of human capital and incorporated the pathways by which human capital and technology mutually interact. The three dimensions of human capital are (1) building and protecting health, education, and skills, as well as income; (2) deploying and utilizing human capital in the labor market and the society at large; and (3) empowering human capital.

Figure 2.1 highlights the complex and multifaceted ways in which technology and human capital interact, as illustrated by the four arrow numbers. First, technology can accelerate the buildup and protection of human capital throughout people's lives, from early years to adolescence and young adulthood to working age and beyond. Examples of the contributions of technology include improved preventive health care (such as vaccines and nutritional information), diagnosis, treatments, telehealth, and delivery of health care through digitalization; tele-education and more effective educational pedagogy through computer-assisted learning, augmented reality, virtual reality, simulations, games, and digital personalization; and improved targeting of social protection programs aimed at poor, vulnerable, and unemployed people, as well as better delivery and payment systems.

Second, technology can improve the performance of other sectors that contribute to human capital such as food and nutrition, clean water and sanitation, electricity, transportation, digital infrastructure, and technologies that help improve the environment

FIGURE 2.1 Relationship between Technology and Human Capital

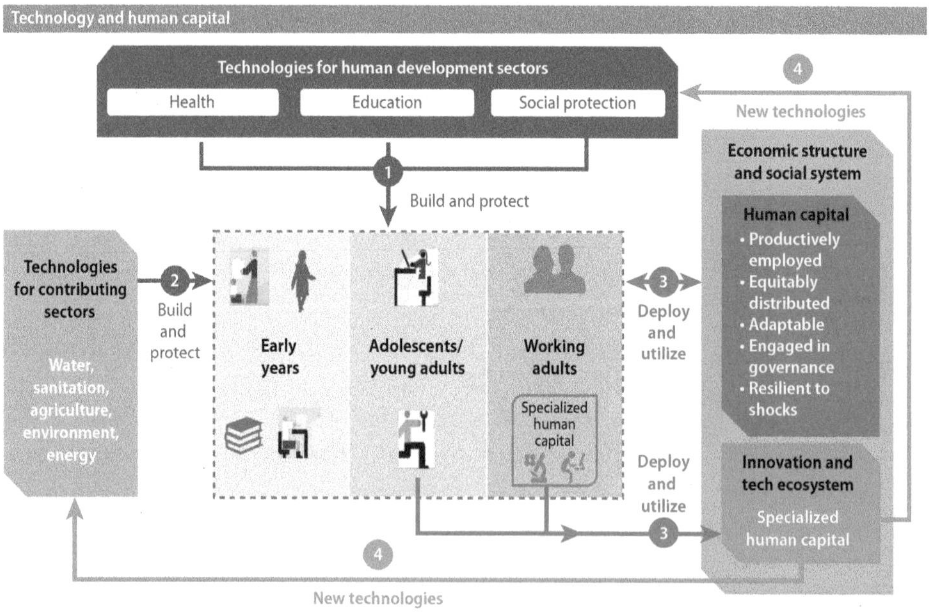

Source: World Bank study team.

by, among other things, reducing air and water pollution, producing noncarbon fuels, and enabling livable cities.

Third, technology also has impacts on the economic structure and social system of a country, which, in turn, affects the sectors that contribute to human capital formation and deployment. For example, changes in the uses of technologies in manufacturing and services alter the demand for the education and skills (such as digital) needed to absorb and deploy new technologies and the need to reskill workers made redundant by new technologies. In addition, technology may have disruptive effects on employment through economic restructuring, which will affect the demand for unemployment insurance and income support for persons in the informal sector who lack unemployment insurance. Meanwhile, the demand for health services may change because of the negative impacts of exposure to new technologies, such as the impact of autonomous vehicles on public health or the impact of automation and robots on workers in factories.

Fourth, another important relationship is between human capital and the creation and, equally important, the adaptation of technology. Specialized human capital (such as technologists and scientists) operating in a well-functioning innovation ecosystem is a critical input in the production of new technology, both for the sectors that directly contribute to producing human capital as well as for those that indirectly contribute to it. This latter point is important. For example, advances in agricultural productivity and improved, low-cost water and sanitation facilities may have big impacts on reducing child malnutrition.

This framework also helps to highlight concerns about inequality and empowerment, which are central to the analysis of human capital. The diffusion and adoption of new technologies tend to favor, particularly in the first round, the more educated, who enjoy greater access to financial and other complementary assets, thereby increasing inequalities. Thus public policy should be directed at how to offset the tendency toward a deepening of inequalities. A second set of concerns relates to the risks to human beings posed by the dual-use nature of many converging technologies, the attendant loss of empowerment and agency, as well as the often unaddressed ethical, moral, and social issues they raise. This line of debate is omnipresent and can be deeply polarizing if left unaddressed. It poses a question: Will people be replaced by machines as rapid advances in automation displace labor as well as knowledge-intensive jobs by means of big data, machine learning, and artificial intelligence? Countries should, then, consider the governance of technology and how to make its use and future deployment transparent and accountable.

The main takeaway from this discussion is that human capital and technologies are closely intertwined at multiple levels. This relationship runs in both directions, whether human capital is a user of new technologies, an input for the effective use of new technologies in the economy, or a contributor to the creation of new technologies. More broadly, human capital is at the center of inclusive development as evidenced by the United Nations' Sustainable Development Goals (SDGs). The *Global Sustainable Development Report* issued by the United Nations places human well-being and capabilities as one of the six key entry points to achieve the SDGs, and it elevates science and technology and individual and collective action as two of the four fundamental levers to achieve the goals (Independent Group of Scientists 2019). Similarly, the International Institute for Applied System Analysis (IIASA) cites human capacity and demography (education, health, aging, labor market, gender, and inequalities) and the digital revolution (artificial intelligence, big data, biotech, nanotech, and autonomous systems) as two of the six transformations that will drive future change as society anticipates "The World in 2050" (TWI2050 2018, 2019).

Within the broad constellation of technologies characterizing the converging technology framework, the study team identified technology metatrends that will affect the South Asia region in the coming years. These metatrends were identified through the scenario exercise described in chapter 7 of the report. These trends are collections of disruptive forces associated with digital and nondigital technologies that exhibit a fairly high degree of certainty about their near- to medium-term impacts, whether positive or negative, on human capital. They reflect the multidisciplinary nature of technologies characteristic of the converging technology revolution and represent overarching global forces that will generate many multidimensional catalytic changes, as opposed to linear or sequential ones. Their positive and negative effects are further discussed in chapter 7 and are included in appendix A of this report.

The four metatrends identified are (1) technologies relevant to the human development sectors and contributing sectors that will improve the access to and the efficiency and quality of health, education, and social protection services, as well as improve human capital outcomes through improvements in agriculture, water, and other sectors; (2) data-driven and human-machine interactions (underpinned by big data, machine learning, and AI) in global and local production processes for goods and services that will affect the level and composition of demand for human capital; (3) inclusive innovation ecosystems that will enable countries to produce, adapt, and diffuse technologies for their local needs and develop scalable business models to accelerate their deployment; and (4) governance mechanisms to address the impact of dual-use technologies on societal norms, agency, empowerment, freedom, and security.

Figure 2.2 lays out a high-level theory of change, linking the pillars of the human capital framework outlined earlier (first column) with technology metatrends (last column). The scope of issues relevant to technology and their potential impacts on human capital outcomes appear in the middle columns. Although there is not a one-to-one mapping of the pillars of the human capital framework and the four metatrends, the color coding indicates where their interaction is the most direct.

FIGURE 2.2 **How Does Technology Enable Human Capital Outcomes and Vice Versa?**

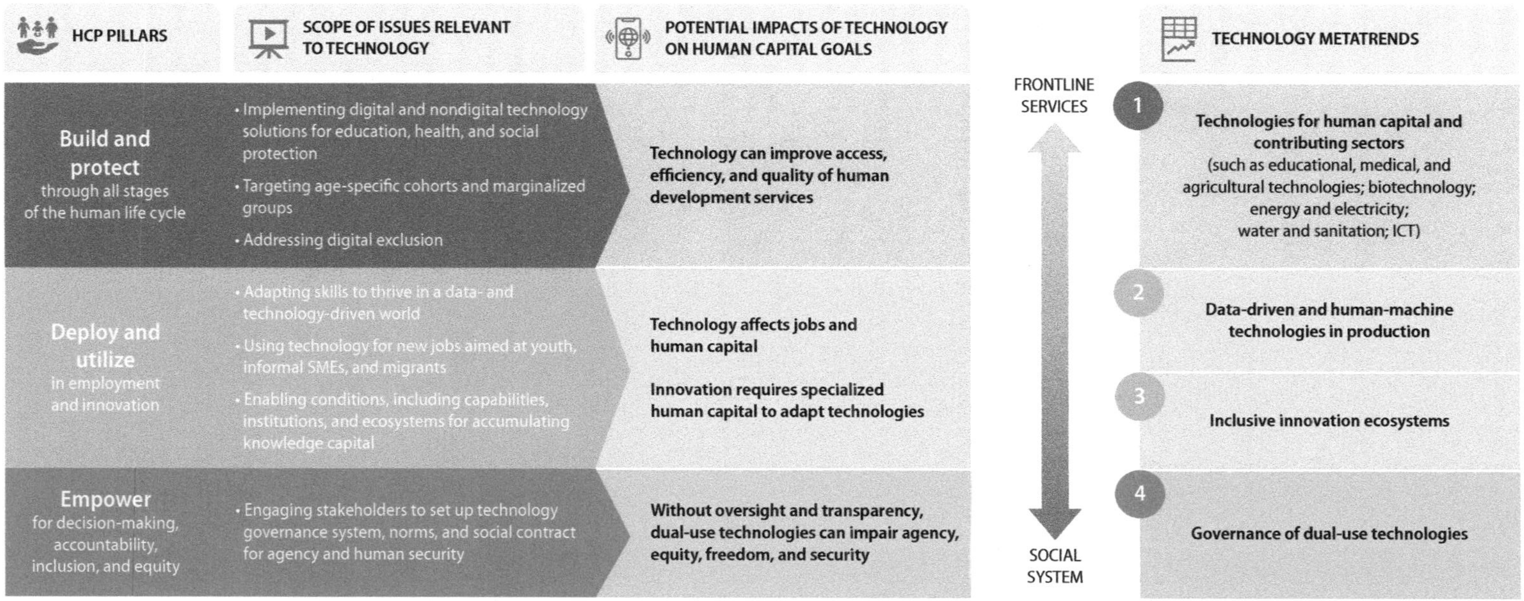

HCP PILLARS

Build and protect
through all stages of the human life cycle

Deploy and utilize
In employment and innovation

Empower
for decision-making, accountability, inclusion, and equity

SCOPE OF ISSUES RELEVANT TO TECHNOLOGY

- Implementing digital and nondigital technology solutions for education, health, and social protection
- Targeting age-specific cohorts and marginalized groups
- Addressing digital exclusion

- Adapting skills to thrive in a data- and technology-driven world
- Using technology for new jobs aimed at youth, informal SMEs, and migrants
- Enabling conditions, including capabilities, institutions, and ecosystems for accumulating knowledge capital

- Engaging stakeholders to set up technology governance system, norms, and social contract for agency and human security

POTENTIAL IMPACTS OF TECHNOLOGY ON HUMAN CAPITAL GOALS

Technology can improve access, efficiency, and quality of human development services

Technology affects jobs and human capital

Innovation requires specialized human capital to adapt technologies

Without oversight and transparency, dual-use technologies can impair agency, equity, freedom, and security

TECHNOLOGY METATRENDS

FRONTLINE SERVICES

1 **Technologies for human capital and contributing sectors** (such as educational, medical, and agricultural technologies; biotechnology; energy and electricity; water and sanitation; ICT)

2 **Data-driven and human-machine technologies in production**

3 **Inclusive innovation ecosystems**

4 **Governance of dual-use technologies**

SOCIAL SYSTEM

Source: World Bank study team.
Note: HCP = Human Capital Project; ICT = information and communications technology; SME = small and medium enterprise.

From Metatrends to Classification of Technologies

The technology metatrends just described summarize the broad disruptive forces that indicate how the world and the region are changing. They are especially useful in undertaking scenario and foresight exercises and in setting strategic directions. For the purposes of analyzing the technologies currently being used or developed in the region in the human development sectors, or in the World Bank's current portfolio of projects, a more granular classification is appropriate.

Within the human development service delivery sectors—health, education, and social protection—that contribute to building and protecting human capital within the study's framework, this study focuses on individual technology products, solutions, or business models, as well as digital platforms, that enable the participation of multiple actors and facilitate the creation of sectoral ecosystems that can transform service delivery (figure 2.3). Individual technology products are further classified into

FIGURE 2.3 Technology Classification Schema and the Human Capital Framework

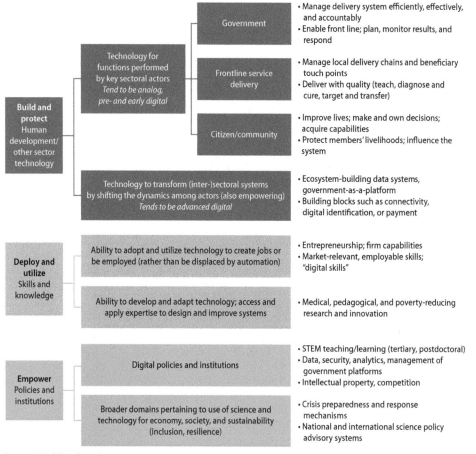

Source: World Bank study team.
Note: STEM = science, technology, engineering, and mathematics.

those used by government, frontline service providers, and citizens or consumers. This broad classification is used to discuss the technology landscape for health, education, and social protection in chapter 3. The analysis of the World Bank's portfolio in chapter 6 also uses this schema, grouping the technology components of the bank's human development projects in the South Asia region by the human capital pillars as well as the broad groups of technologies. The upstream discovery (or adaptation) of new technologies or scientific or engineering methods is an important aspect of a country's future use of technology. The capacity to discover new technologies and scientific or engineering methods is examined in conjunction with the discussion of innovation systems in chapter 4. The broader policies and institutions that support the safe, equitable development and use of technologies are critical to the empowerment of human capital. Because of the importance of data and artificial intelligence in the converging technology revolution, these subjects are addressed in chapter 5.

This schema, while useful for the purposes of this study, is naturally somewhat arbitrary. As discussed in the literature, these technologies can be classified in many ways: (1) broad technology areas, such as digital, biological, and materials; (2) function, such as whether they reduce costs or improve delivery of services; and (3) user group, such as individuals, businesses, or organizations. For the health sector, the World Health Organization (WHO) has proposed a classification that, for digital health technology alone, has nearly 90 technologies segmented by four principal actors: clients, health providers, health system managers, and data services. For the education sector, Nolon, a private education company, has proposed a classification of some 60 technologies that mix technologies and users.

Priorities for Human Capital in South Asia

The World Bank's South Asia Human Capital Plan diagnoses the critical drivers of the areas in which the World Bank will prioritize its investments and engagement between 2020 and 2025. Three key drivers limit human capital outcomes in the region: (1) the poor quality and poor effectiveness of services, both of which undermine the impact of public investments; (2) the multiple inequalities that lead to large segments of the population being left behind; and (3) people facing increasing vulnerability to a range of shocks and risks. The strategic priorities for accelerating human capital are smarter and higher-quality investments; inclusion and empowerment, especially of adolescent girls and women; insuring and preparing for potential shocks; and innovating through data, technology, and multisector action.

The plan also identifies critical human capital challenges in the region such as the lifetime burdens of stunting and limited learning in the early years, which are associated with and are markers of deep socioeconomic inequality. Both need to be addressed urgently because they have persisted for decades. Another priority is reducing mortality and morbidity from noncommunicable diseases and water and air pollution. Addressing these challenges requires improving the quality of services, especially last-mile service delivery for the poor and the vulnerable, by increasing spending, efficiency, and accountability using both the public and private sectors.

The plan focuses as well on developing twenty-first-century skills and employment opportunities for youth, which are essential to reducing intergenerational inequality, as well as safeguarding displaced human capital.

Finally, the plan highlights the fact that the empowerment agenda for human capital in South Asia, which also aligns with the third pillar of this study's framework, is centered on girls, women, and marginalized communities, who are subject to deprivation, discrimination, violence, and exclusion in many forms.

The challenge therefore is to harness the converging technology revolution to not only improve service delivery ("building and protecting" human capital), but also to ensure better jobs ("deploying and utilizing" human capital) and reduce inequalities, strengthen inclusion, and enable empowerment ("empowering" human capital). Without adequate regulation and intentional public policies, the latter two objectives could be easily overlooked or undermined because of some of the inherent features of these technologies. This study also highlights the need to build the innovation capacity of countries so they are able to locally adapt, diffuse, and develop technologies for a resilient future.

Perspectives from the Region: Country Expert Interviews

Discussions with local actors in the region revealed other perspectives on the priorities for human capital and the use of technology in South Asia. A series of virtual interviews were held with technology experts in Kerala (India), Nepal, and Pakistan.[6] The interviews were extremely helpful in understanding the situation on the ground, including political economy issues and geopolitical rivalries that affect technology choices, and in eliciting suggestions for what the World Bank could do in the future. Table 2.1 is a summary of the responses by each group.

A key insight from these interviews is that those at the local level broadly recognize the qualitative shift in the uptake of technology arising from COVID-19. And it is widely understood that such disasters can and will happen again, thereby underscoring the urgency of anchoring disaster preparedness and resilience at the community level. Such a step would include building capabilities for community innovations and adaptation of solutions to local needs ("resilience by design") and strengthening the capacity of local government. Resilience to future shocks requires greater digital collaboration between the public and private sectors and digital leadership capabilities within government to leverage digital technologies. There is a potential for quick wins in health, education, and social protection, including assessing requirements for critical medical and food supplies, rethinking the traditional schoolhouse delivery mode with roving teachers and mobile learning labs, and delivery of social assistance through financial

TABLE 2.1 **Summary of Interview Responses: Kerala (India), Nepal, and Pakistan**

	Kerala: Long-term investment in STI, community capacity, and engagement	Nepal: Many local innovations; private sector diaspora willing to engage	Pakistan: Under crisis, government willing to adopt joint solutions with the private sector
What has changed as a result of COVID-19?	• Disaster preparedness anchored over decades in science clubs in schools, universal literacy, land reforms, decentralized planning with local governments, and active citizen involvement in saving rainforests • Relevant experience in using technology during the floods of 2018 and Nipah virus • Mass volunteerism; spontaneous demand from society for STI solutions; widespread use of open-source programming • Local resilience considered important, such as nongrid solutions for water, electricity, and waste management, especially as there is no centralized legacy system	• Technology adoption on user side; diverse microlevel innovations; ongoing engagement with key stakeholder groups (such as school administrators and health care workers) to address their concerns • Strong demands by private sector and diaspora to implement national digital strategy in coordination with government	• Launch of educational TV with interactive SMS and private educational content • Telemedicine platforms and local production of PPE but lack of entrepreneurial drive and solutions orientation by private sector
What has the greatest potential for quick wins?	• Open-source platforms for developing innovative products; FABLABs (maker spaces) linked to enterprises • Colleges and universities operating as incubators, focusing on finding solutions to local problems	• Use fintech and point-of-service network to improve delivery of social transfers and lay groundwork for digital ID registration • Ramp up tech training for teachers; pilot "roving teachers" and mobile learning centers to reach local communities • Expand Nepal's narrative beyond tourism to demonstrate the country's digital potential through incubation centers and support for start-ups • Build an up-to-date health profile of Nepal's provinces to address priority needs, including through telemedicine	• High performance of national ID system; potential for cross-platform integration of human development service delivery • Rethinking of traditional schoolhouse education delivery model • Elimination of fifth-grade examination to reduce stigma of school choice • Solutions for reaching a high level of adult illiteracy; incentives for national campaigns (such as offering smartphones as rewards upon successful completion)

(Table continues on next page)

TABLE 2.1 Summary of interview responses: Kerala (India), Nepal, and Pakistan (continued)

	Kerala: Long-term investment in STI, community capacity, and engagement	Nepal: Many local innovations; private sector diaspora willing to engage	Pakistan: Under crisis, government willing to adopt joint solutions with the private sector
What are the risks posed by new technologies?.	• Significant number will remain poor and gender and caste inequalities will remain strong • Big social gap in education, traditionally the equalizer, is emerging • Risk stratification approach poses complex cyber risks and raises question of data proportionality (such as what happens to data collected on children?)	• Growing digital divide, glamorizing of online education; growing number of dropouts • Lack of legal, cyber, and privacy safeguards	• Education sector is highly segmented; private education sector may capture the benefits of technological progress • Poor educational performance remains a national emergency—10 years after being declared an emergency! • Government of Pakistan is unable to coordinate and implement technology initiatives; local government capacity is nonexistent • COVID-19 may continue for five years, thereby deepening inequality • Disinformation is a BIG challenge
What are examples of promising grassroots innovations?	• "Taking the state to the doorsteps of vulnerable" • Local science clubs linked to international networks—for example, IEEE's 4,000 members were able to come up with local innovations faster than major companies • Kerala's investments in techno parks, start-ups, e-governance, and STI institutions are enabling broad-based innovation	• Returning migrants and diaspora with IT/AI expertise present opportunities for digital innovation • Small-scale innovations in education, health, and e-commerce are being made, but require funding and mentorship	• Creation of STEM program for age group 13–17 years • Local accelerator programs (such as Life Straw water purification system)

(Table continues on next page)

TABLE 2.1 **Summary of interview responses: Kerala (India), Nepal, and Pakistan**
(continued)

	Kerala: Long-term investment in STI, community capacity, and engagement	Nepal: Many local innovations; private sector diaspora willing to engage	Pakistan: Under crisis, government willing to adopt joint solutions with the private sector
What can help build resilience for future livelihoods?	• Building resilience at the micro/community level (lesson from floods) • Adapting solutions to local needs inspires "resilience by design" • Moving from central grid to local grids built for resilience • Receiving World Bank financing through its projects for widespread, decentralized adoption of technology at the local level (such as from development to pilot plants)	• Jointly piloting and expanding social registry across human development services with a cohort of proactive local governments • Stockpiling critical medical supplies in preparation for future shocks and seizing opportunities to achieve UHC • Improving collaboration between the public and private sectors and encouraging government's digital leadership • Strengthening donor coordination around digital Nepal • Using technology to upgrade disaster response system and inform communities	• Building up vertical/specialist skills in government of Pakistan

Source: World Bank study team.
Note: AI = artificial intelligence; ID = identification; IEEE = Institute of Electrical and Electronics Engineers; IT = information technology; PPE = personal protective equipment; SMS = short message service; STEM = science, technology, engineering, and mathematics; STI = science, technology, and innovation; UHC = universal health care.

technology (fintech) or mobile phones. Citing several examples of promising grassroots innovations, interview participants noted that such innovations typically do not receive adequate, if any, funding or support from donor-funded projects. They also pointed out that the greatest risks are growing inequality and cyber threats, as well as extensive data appropriation without consent (such as from children). Later chapters will expand on these risks.

Summary

Converging technologies are variable combinations of information and bio-, nano-, and cognitive technologies that are amplified through artificial intelligence, big data, computing power, and connectivity. The promise of converging technologies for improving the targeting, customization, and delivery of services is accompanied by downside risks associated with unintended consequences and, in some cases, malicious intent. As converging technologies affect ever more aspects of daily lives, it is critically important that people endeavor to understand and shape the fundamental ethical, moral, and governance-related issues surrounding the use of technologies and their relationship to human capital lest they be subjugated to decisions made elsewhere.

The ultimate value added of the human capital framework is to identify priorities for action in the South Asia region. For the "building and protecting" dimension, these include, for example, focusing on digital access and technology-enabled services to address inequality in health and education outcomes, persistent stunting, and the absence of social insurance coverage. Over the next decade, the "deployment and utilization" dimension of human capital will assume great urgency to recover from the pandemic, to respond to the need to create a large number of jobs in the economy, and to create resilience in the face of climate shocks and environmental degradation that are besetting the subcontinent. And the "empowerment" dimension will warrant special attention because it sets the conditions for transparency, accountability, data protection, and safeguards against the adverse actions made possible by dual-use technologies. The three chapters that follow are structured, respectively, around these three dimensions of the human capital framework developed in this chapter.

Notes

1. Technological revolutions are periods in which there is strong interconnectedness and interdependence of technologies that transform the economy and society. Such periods have been called techno-economic paradigms (Perez 2002). Examples of earlier technological revolutions are the Agricultural Revolution, the Industrial Revolution, and the more informally termed Information Revolution.

2. Many may argue that the Fourth Industrial Revolution is still part of the Information Revolution or its extension. However, many qualitatively different elements justify distinguishing the converging technology revolution from the Information Revolution.

3. The Human Capital Project is a global effort facilitated by the World Bank to accelerate more and better investments in people for greater equity and economic growth (https://www.worldbank.org/en/publication/human-capital).

4. As a review by the World Bank's Independent Evaluation Group indicates, various terms are used within the World Bank Group, including *disruptive technologies, digital technologies, and transformative technologies* (World Bank 2021). The study team uses the term *converging technologies* to highlight the specific aspects of technologies that are affecting human capital outcomes.

5. The HCI conveys the productivity of the next generation of workers in relation to a benchmark of complete education and full health. It is made up of five indicators: (1) the probability of survival to age five; (2) a child's expected years of schooling; (3) harmonized test scores as a measure of quality of learning; (4) adult survival rate (share of 15-year-olds who will survive to age 60); and (5) the proportion of children who are not stunted.

6. These experts are listed in the acknowledgments of this report. Although the insights from these interviews constitute findings of this study, they are included in this chapter because they help to set the stage for the analysis and discussions in the rest of the report.

References

Independent Group of Scientists. 2019. *Global Sustainable Development Report: The Future Is Now*. New York: United Nations.

Perez, Carlota. 2002. *Technological Revolutions and Financial Capital: The Dynamics of Bubbles and Golden Ages*. Cheltenham, UK: Edward Elgar.

Schwab, K. 2016. "Fourth Industrial Revolution: What It Means, How to Respond." World Economic Forum, Geneva. https://www.weforum.org/agenda/2016/01/the-fourth-industrial -revolution-what-it-means-and-how-to-respond/.

TWI2050 (The World in 2050). 2018. *Transformations to Achieve the Sustainable Development Goals. Report Prepared by The World in 2050 Initiative*. International Institute for Applied Systems Analysis (IIASA), Laxenburg, Austria. http://pure.iiasa.ac.at/15347.

TWI2050 (The World in 2050). 2019. *The Digital Revolution and Sustainable Development: Opportunities and Challenges. Report Prepared by The World in 2050 Initiative*. International Institute for Applied Systems Analysis (IIASA), Laxenburg, Austria. http://pure.iiasa.ac .at/15913/.

World Bank. 2016. "Human Capital Project: Year 2 Progress Report." World Bank, Washington, DC. https://documents.worldbank.org/en/publication/documents-reports/documentdetail /483491602866728302/human-capital-project-year-2-progress-report?cid=GGH_e _hcpexternal_en_ext.

World Bank. 2019. "The Human Capital Project: An Update." Presentation prepared by World Bank Group for October 19, 2019, Development Committee Meeting, Washington, DC.

World Bank. 2021. *Mobilizing Technology for Development: An Assessment of World Bank Group Preparedness*. Independent Evaluation Group. Washington, DC: World Bank. https:// openknowledge.worldbank.org/handle/10986/34517.

CHAPTER 3

Building and Protecting Human Capital: The Technology Landscape for Service Delivery in South Asia

Introduction

This chapter describes the landscape of technologies being deployed in South Asia in health, education, and social protection to help build and protect human capital, following the framework outlined in figure 2.1 in chapter 2. Technologies being rapidly deployed in sectors that indirectly contribute to human capital (such as agriculture and sanitation), while extremely important, are not covered.

The opportunities from new digital, data-driven, and converging technologies arise from their ability to reach the unreached, improve quality and efficiency, and enable personalization and differentiation according to need. In the public sector, the potential also lies in reimagining citizen-centric service delivery models and fostering data-driven decision-making. Yet these opportunities are limited by inequalities in digital access, not only in broadband connectivity and in the availability of electricity, but also in devices, people's digital skills, and local content, which risk deepening the human capital inequalities in South Asia.

The chapter begins by discussing the opportunities and the characteristics of technology markets in human development in South Asia, followed by an overview of the inequalities in digital access that limit these opportunities. It then moves to a summary of the technology landscape in health, education, and social protection in South Asia, including a brief look at two case studies that highlight applications of converging technologies in the human development sectors, and the use of technology in the initial phases of the response to the first wave of the COVID-19 pandemic.

The chapter concludes with opportunities for data-driven decision-making to improve planning, implementation, monitoring, and accountability of human development service delivery.

Opportunities for Improving Service Delivery in Health, Education, and Social Protection

Converging technologies have not yet penetrated the human development sectors to the same extent as the agriculture and industry sectors. Within the human development sectors, innovation in service delivery is being driven mainly by information technology applications. The principal trend within each sector is the shift from individual nondigital and digital technology tools to digital platforms and data-based technologies using machine learning and artificial intelligence (AI). Within the latter, an important distinction can be made between individual digital technology tools, which are mostly about one-way interactions; digital platforms, which combine interactions across multiple agents; and systems enabled by big data, machine learning, and AI that seek to extract value from data and produce knowledge that can then be reused. Although the distinction between the last two categories is somewhat arbitrary (because the data generated by platforms can also be exploited through machine learning and AI), the idea is to distinguish between digital technologies that allow coordination and delivery of existing knowledge and content and those that generate new knowledge used for commercial or social purposes. In summary, these technologies enable innovations across the entire service delivery value chain that can generate significant benefits or cost savings.

TECHNOLOGY GROUPS

Digital public platforms enable governments to interact virtually with frontline service providers and beneficiaries through web- and mobile-based sites, applications, and software that provide the interfaces for these interactions. First-generation, citizen-facing platforms were developed to provide information (for example, to announce public examination results), but they are increasingly moving to two-way transactional service delivery. As noted in the description of the World Bank's Digital Economy for Africa diagnostic tool, "Digital public platforms require digitalized systems and processes, shared and interoperable resources, interfaces for internal and external users, digital authentication capability, and online trust. These functionalities are provided through various core components, including digital identification and trust services, interoperability layers, and shared services" (World Bank 2020, 32).

Client-facing services—that is, those in which employees interact directly with customers—can be offered through various digital channels, such as online portals, mobile phones, and social media. Digital public platforms can help improve the cost and efficiency of service delivery and offer beneficiaries ease and convenience. Digital public platforms can be built and run directly by the government or in partnership with private firms. Although the latter approach helps to compensate for the lack of technical capacity within governments, other capabilities are required to define, procure, and manage these partnerships to avoid vendor lock-in and so on.

A special category of digital public platforms for the human development sectors is dynamic social registries that connect people to a range of public services, including social protection, health, education, utilities, and financial inclusion. They help to expand coverage of social services and prioritize the poorest people or targeted groups, while also allowing for scale-up and targeting assistance in the face of crises, disasters, and other shocks. Furthermore, these platforms, by linking to a unique identification (ID), can help to reduce costs associated with inclusion and exclusion errors.

Within human development sectors, the converging technology revolution is most advanced and prominent in the health sector, where it can have immediate impacts on human capital accumulation and welfare. Technology convergence is apparent in the entire global value chain, including digitalization of life science in drug discovery; nonimaging diagnostics based on genomics; nanotechnology in drug delivery; additive manufacturing for medical devices; use of drones for medical supply delivery; the Internet of Medical Things; use of robots in surgery; virtual and augmented reality for remote surgery and training of doctors, nurses, and technicians; and geospatial and epidemiological modeling for disease prediction and surveillance. Portable genomics sequencers allow for diagnosis of diseases in real time in remote areas. AI software, together with sensors and cameras, can turn mobile phones into sophisticated diagnostic tools, which can be used for digital microscopy, cytometry, immunoassay tests, monitoring vital signs, and detection of malaria or cervical cancer. AI has the potential to make medical devices more reliable, and it is helping to decentralize care that can be delivered to those without access to health facilities. Furthermore, genomics, although only a subset of nonimaging diagnostics, is a critical data source for enabling the transition to personalized medicine, especially when combined with AI and other tools.

An equally important shift is occurring toward a whole-of-government approach to digitalization for the delivery of public services and the use of data-driven decision-making in the design and implementation of policies and services and to monitor performance and measure impact. Sophisticated systems of data-driven decision-making integrate data from multiple sources, such as geospatial and remote sensing data and household level data, to inform policy making and implementation.

Table 3.1 summarizes the broad groups of technologies and their pathways to improving service delivery for enhancement of human capital.

TABLE 3.1 Technologies for Improving Service Delivery for Building and Protecting Human Capital

Technology groups and examples	Pathways to improving service delivery for human capital

1. Individual nondigital technology tools
  • Education tools
  • Drugs and vaccines, medical devices
  • New medical treatments
	• Offer potential to lower costs and improve efficiency or effectiveness of service delivery

2. Individual digital technologies
  • First-generation ICT-based management information systems for health, education, and social protection
  • Cellphones and mobile applications for education, health, and nutrition
  • Drones for remote sensing and delivery of medical and disaster relief supplies
  • Satellite imaging for monitoring agriculture, weather, disasters, and movements of people
  • Blockchain to ensure security in digital identity and payment systems
  • 3-D printing for medical supplies, prosthetics, teaching modules, and so on
  • Augmented reality and virtual reality training, information, and remote learning experiences to aid students and to train workers, including doctors and other health personnel, to use new technologies
  • Online health, nutrition, and educational content
	• Improve efficiency in delivery and content of health, nutrition, education, and social protection services
	• Help to identify potential beneficiary pool to provide services to populations that have been marginalized, subjected to discrimination, or are otherwise at risk
	• Allow more effective communication with citizens

3. Digital platforms
  • Social registries integrated with other systems, including digital payment systems for social protection services
  • Online education portals with learning resources and open course enrollment to enable teachers and students to access multiple education resources
  • E-health (digital health)
  • Matching medical supplies with needs at clinical care facilities and patients with available health services
  • Job-matching platforms
	• Reduce asymmetry of information and the costs of accessing information and allow interaction between consumers and providers of goods and services
	• Enable inclusion by giving marginalized groups and communities access to services in their homes
	• Foster innovation through new products and services based on monetizing data, including health and education

4. Data-driven, personalized applications of converging technologies in health and education
  • AI-driven personalized and adaptive learning, including personalized massive online open courses
  • Assessment of students' performance
  • AI-driven personalized health services
  • Affective computing
  • Malnutrition measurement and monitoring at the individual level
  • Human–machine interaction and augmentation
	• Augment the effectiveness of a service through self-learning technologies, using instant feedback and information from beneficiaries
	• Help traditionally excluded marginalized and disadvantaged groups by addressing their specific needs
	• Augment individual capacities

5. Data-driven decision-making technologies, combining big data, AI, geospatial, and related technologies
  • Big data analysis and remote sensing, using mobile phone records to track people's movements
  • Early warning systems for weather or epidemics
	• Improve accuracy of decision-making and predictions based on available data
	• Support the use of data optimization and predictive intelligence to target services and protect human capital during disasters and crises
	• Combine knowledge arising from datasets for multiple sectors to improve planning and monitoring

Source: World Bank study team.
Note: AI = artificial intelligence; ICT = information and communications technology.

TECHNOLOGY MARKETS AND ADOPTION OF TECHNOLOGY

In assessing opportunities for the use of technology in the human development sectors, it is important to understand the structure of the market and global technology trends. The providers and disseminators of technology are largely private firms operating through the market system. Many times, they do not have an incentive to develop the kinds of products or services needed to reach poor populations because the poor are not considered—often based on faulty assumptions and lack of knowledge—to be profitable market segments. On the demand side, purchasers are governments, private health care and education providers, and individuals. Governments often lack the technical capacity to evaluate technology options and are slow to adjust the policy and regulatory frameworks that inhibit the adoption of technology (or enable it without adequate safeguards for citizens).

There are also significant differences among the three sectors. Health and education services are provided by both governments and the private sector, with significant differences in consumer segments. Social protection services are largely offered by governments to individuals and communities, the majority of them poor. Some nonprofit organizations may also provide certain types of social assistance, but they are usually relatively small in relation to government operations. Social protection systems mainly use two technologies: digital platforms (which may be developed or operated in partnership with private partners) and payment systems. Technologies tend to be broadly similar across countries, although legal and regulatory frameworks will affect the extent to which interoperability and shared services can be put into effect.

The distinguishing features of health technology, however, are that, on the one hand, the pace of innovation and the scope of technology products and services are extremely varied, and, on the other, broadly similar technologies are deployed across the world for prevention, diagnosis, treatment, and alleviation of disease or injury. Private investment and knowledge flows across the globe are driving the development of new technologies in the health sector. The standardized training and certification of medical professionals worldwide—even if quality varies—and the international professional and private health provider networks mean that knowledge exchanges across borders are fluid, with greater opportunities for health care providers and medical professionals to learn how to use new technologies.

Private firms and health care providers in developing countries import technologies for deployment in private hospitals serving wealthier households, accentuating already existing health inequalities. One example is the rapid diffusion of robotic surgery technology in large private hospitals in developing countries. The proliferation of diagnostic tools aimed directly at the consumer also benefits wealthier households.

In the education sector, by contrast, technology markets were much more localized until the recent advent of digital technologies, platforms, and AI tools. The use of local languages, the country-specific nature of curricula (with different local curricula), a highly labor-intensive delivery model, and the lower levels of technical skills of most teachers imply a slower rate of technology adaptation and diffusion. The advent of digital technologies and AI-powered tools have altered this to a certain extent as global education technology (edtech) giants penetrate markets, and customization to local markets is increasingly made possible. Nevertheless, the deployment of technology tools tends to serve the higher-end consumers and elite educational institutions because of the high cost of services and connectivity and bandwidth requirements.

Digital public platforms are a powerful means of reducing these inequalities. But beyond this, local innovations are needed to adapt technologies to local needs and provide them at an affordable cost. Such innovations are often constrained because few local firms have the technological capabilities, technical skills, capital, and incentives to undertake such innovations. Governments can thus play an important role in fostering and incentivizing the local private sector to adapt technologies for building human capital.

Unequal Digital Access in South Asia: Barriers to Equitable Deployment of Technology

The tremendous opportunities provided by digital and converging technologies, including public platforms, cannot be exploited unless there is equitable digital access. Digital access should be understood in a broad sense, spanning affordable broadband connectivity and devices, availability of local language content, and the basic digital skills needed to utilize the technology. Access to electricity (or alternative energy sources) to power devices and access broadband must also be part of the solution. An important agenda for developing human capital in South Asia is therefore to advocate for bridging these divides to enable disadvantaged groups to benefit from service delivery and participate in the digital economy. In the short run, however, the most critical step is to ensure that the design of digital solutions, including the use of digital public platforms, takes into account adaptation of the final delivery mechanisms to the conditions faced by beneficiaries, such as lack of smartphones, lack of access to fixed broadband, lack of electricity, and limited digital skills and digital literacy.

Gaps in last-mile connectivity and affordability. The "last mile," an expression referring to the last segment of the journey of broadband connectivity to its recipient, is in effect the first mile in the delivery of human development services. The South Asia region is home to the largest number of people without an internet connection—nearly a billion people out of a worldwide total of 3.2 billion. Data supplied by mobile phone operators reveal the differences in broadband speed and coverage across countries. In Afghanistan, for example, 90 percent of the population has access only to 2G networks, and only 2 percent have access to 4G networks. In Sri Lanka, 4G networks cover 90 percent of the population. In all countries, internet access is mostly through mobile phones, which limits the delivery at the household or individual level of education services for remote teaching or self-learning and of health services that require prolonged interaction, among other functionalities. Barring India, other countries have limited fixed-line infrastructure, which affects the fixed broadband penetration rate and reliable high-speed connectivity, even for schools, health centers, and hospitals. International bandwidth usage is less than 10 kilobits per second per user.

The average price of broadband in South Asia is less than 2 percent of the per capita gross national income (GNI), the affordability benchmark set by the UN Broadband Commission. The exception is Nepal where the cost is closer to 8 percent of GNI (and data are not available for Afghanistan). However, beyond these average figures, the picture is much less sanguine. For example, reportedly the bottom 60 percent of the population in Sri Lanka cannot afford broadband, even though in this country the average broadband price is one of the lowest in the world (World Bank, forthcoming).

Device availability. The type of digital device affects the quality of the digital service as well as its use for different purposes. The limited data on devices available for countries in South Asia reveal the deep gaps limiting the uptake of digital services. In Sri Lanka, about 52 percent of households with children under 18 did not have access to smartphones, tablets, or computers in 2018.

Inequalities between households and differential access within households. Data from household surveys is relatively sparse regarding access to mobile phones, the internet, and devices. According to the National Sample Survey of 2017–18, only 24 percent of households in India had access to the internet, and rural availability was just 15 percent, compared with 42 percent of households in urban areas. A recent cross-country household survey of selected countries in South Asia and Africa uncovered deep gaps (LIRNEAsia 2019). Access to desktop or laptop computers is less than 10 percent in the five South Asian countries surveyed (figure 3.1, panel a). The gender gap in mobile phone ownership is the highest in India, followed by Bangladesh and Pakistan (panel b). In fact, the report indicates that gender gaps in South Asia are higher than in the Sub-Saharan African countries surveyed. Similar gaps by income level (panel c) are especially pronounced in Pakistan. Internet awareness is usually less than 50 percent, except in Sri Lanka, and usage is below 20 percent, except in Nepal and Sri Lanka, where one-third of the population reports regular usage (panel d). The low usage rates may reflect multiple factors, including limited device availability at the household level, lack of digital skills, scarcity of local language content, and limited adoption of digital services that familiarize users with the internet.

Local content. The proportion of content hosted overseas and accessed by users in South Asia is not known. This content must cross expensive, and at times unstable, international links, which affects the use of these resources. India is likely an exception because of its more advanced digital infrastructure, including its in-country data centers.

Moreover, the availability of locally relevant content in local languages is an extremely important issue. Although most South Asian countries use vernacular languages in analog form for government services, including health and education, the availability of local

FIGURE 3.1 **Mobile Phone and Desktop/Laptop Ownership by Income and Gender and Internet Awareness and Usage: Selected Countries, South Asia**

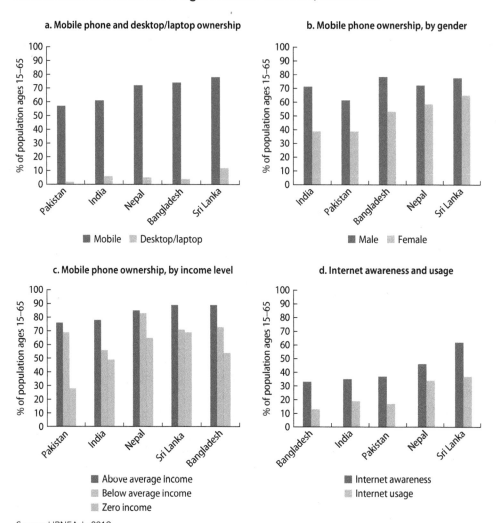

Source: LIRNEAsia 2019.

Note: Individual data points may deviate from the LIRNEAsia survey, reflecting different reporting periods, survey methods, and indicator definitions. For example, according to the 2019 Multiple Indicator Cluster Survey for Nepal conducted by the Central Bureau of Statistics and the United Nations Children's Fund (UNICEF), 35.3 percent of women and 55.5 percent of men reported they had used the internet at least once a week during the last three months, averaging around 45 percent for the population ages 15–65.

language digital content for education is still relatively limited. Linguistic minorities are severely disadvantaged as well. Locally produced content and local hosting is important for developing the local internet ecosystem and the adoption of digital tools. On the flip side, digital tools in vernacular languages, including "minority" languages, could be used to teach literacy as well as digital skills.

These deep gaps across regional, socioeconomic, gender, and language groups affect the use of technology to address inclusion and inequality in human capital. They became apparent in the COVID-19 pandemic, as discussed later in this chapter.

Technology Landscape in Health, Education, and Social Protection in South Asia

Although there is a significant level of public provision and financing of primary and basic education in South Asia, private providers and household financing dominate the higher levels of education (that is, secondary and higher education and skills training). The private sector is also the predominant provider of health care. In fact, the region's public sector financing of health care as a share of the gross domestic product (GDP) is among the lowest in the world. Private sector participation in both education and health is highly differentiated, with high-quality educational institutions and hospitals accessed by the wealthiest households and with low-cost and poor-quality service providers for the urban poor and those living in periurban and wealthier rural areas. These characteristics of the provision and financing of the deployment of technology in different segments of the education sector highlight the risk of deepening inequalities in education.

The technology landscape in South Asia in the human development sector is vibrant with multiple suppliers in the private sector and the deployment of a host of technologies, especially in the private sector. The domestic technological capacity varies substantially, with India being the clear leader. Public sector capabilities of countries to deploy technology also vary, but most applications in service delivery in all three sectors involve digital platforms. Figure 3.2 summarizes the applications of technologies to human development in the three sectors in South Asia. The rest of this chapter describes the trends in health, education, and social protection and two case studies in the use of converging technologies.

HEALTH

Health technology markets in South Asia are dynamic, with India the regional hub connecting global and adjacent local markets. Indicative analysis shows that the health and technology–related industry in India may be as large as those in China and Japan, may be equivalent to 20–40 percent of the industry in the European Union, and is larger than that of the rest of the South Asia region by a factor of 30 or more. In South Asia, health technology and digital health solutions are supplied by five nongovernmental segments:

- **Global pharmaceutical companies and medical device manufacturers** (such as Pfizer, Roche, Novartis, Johnson & Johnson, Medtronic, GE, and Omron) have

FIGURE 3.2 **Human Development Technology Landscapes in Health, Education, and Social Protection Sectors, South Asia**

		Health	Education	Social protection
What technologies exist, and how can one classify them?		• Medtech is broad (drugs, vaccines, devices) • Focused on digital health by actors (government, payer, provider, patient, data services)	• For K-12, edtech by actors (government, schools, teachers, learners) • NREN for higher education institutions	• Foundational systems for social protection and other services—unique ID, social registry, and digital payment
What is under way in South Asia?	**Private sector**	• Global (Big Tech), regional (largely from India), and local, relatively consolidated due to higher capital requirements • Expected growth, but market not yet ready (COVID-19 likely changing prospects)	• Fragmented locally (e.g., 4,000 edtech startups in India), due to language, culture, and institutions; a few global edtech companies • Edtech high growth, driven by B2C	• Local (such as vendor to government, except private payment networks where government is slow)
	Governments — To enable opportunities	• Strong push through digital health strategies, EMR guidance, and data interoperability systems • Frontline capacities likely a challenge	• Some countries have been active, while frontline capacities tend to be inadequate (COVID-19 forced low-tech solutions)	• Consolidation challenges (such as 440 registries in India and 140 in Bangladesh) • Implementation challenges (such as Nepal counterpart restructuring)
	Governments — To mitigate risks	• Rising awareness on potential data risks • Not an easy legislative environment (such as India's center-state coordination—for example, Aadhaar and/or new ID)	• Conventional efforts on digital divides • Little awareness and few debates on data risks	• Widening debates on privacy, data protection, and civil rights • Ongoing legislation in some countries and states (including through World Bank pipeline DPL)
	Development partners	• Multilateral and bilateral partners and philanthropies increasingly active in norms, standards, benchmarking, operational toolkits, and data collaboratives	• A few visible supporters (such as DFID, but facing a reorganization)	• Strategization and advocacy among bilateral donor agencies (DFID, GIZ, Australia, USAID)

Source: World Bank study team.

Note: The arrows indicate the trend over time. B2C = business to consumer; DFID = Department for International Development (UK); DPL = Development Policy Loan; edtech = education technology; EMR = electronic medical record; GIZ = Deutsche Gesellschaft für Internationale Zusammenarbeit; ID = identification; medtech = medical technology; NREN = national research and education network; USAID = US Agency for International Development.

research and development (R&D) centers, sales, and maintenance locations or local subsidiaries in India and serve neighboring markets.

- **Local pharmaceutical industries** in India serve not only regional but also global markets as the leading exporters of generic drugs and vaccines. However, even in India the local medical device industry is relatively underdeveloped. High-end technology devices are largely imported. Local industry focuses on the less technologically complex products (such as consumables and implants) for both domestic sales and exports. It has not introduced significant innovations to meet local health needs because, among other things, it lacks a "component manufacturing ecosystem and skills base to support local industry" (Deloitte and Healthcare Federation of India 2016).

- **Global Big Tech** often invests in *local technology start-ups*, facilitating foreign direct R&D investments as well as technology and talent flows. Google, Microsoft, and Tencent account for 70 percent of big technology global investments in 15 categories of digital health. Of the largest group—Google's 57 digital health portfolio companies—at least a quarter are active in India, of which one-third has expanded into other South Asian countries. Bangladesh is clearly the next destination. Amazon, Apple, IBM, GE, Cisco, Samsung, and Alibaba have local partners (public and private hospital chains, online pharmacies, and insurance companies) to conduct medical AI research, support clinical decisions, and expand telemedicine or directly serve customers.

- **Local technology conglomerates** (such as HCL, Infosys, Tata, Reliance, and Wipro), including through partnerships with global players (such as Microsoft, GE, and Oracle), provide health system integration for hospitals and governments, digital clinical trials, fitness service platforms, and applications for consumers.

- **Social enterprises and corporate social responsibility (CSR) activities** involving global and local actors from the private sector and civil society are supported by corporate funding under India's Companies Act 2013.[1] Prominent global actors include Microsoft (AI for Good), Cisco (digital hospital), HP (eHealth centers), and Samsung (smart health). Global Innovation Exchange, a global innovation platform, supports more than 600 social enterprises in South Asia in the health field.

- **Donor-funded activities** are supported by multilateral and bilateral development partners or philanthropies, and they typically are delivered by local nonprofit organizations or in partnership with the private sector (and partially overlapping with development partners or philanthropies) with an explicit focus on delivering technology-enabled solutions for key development priorities. For maternal, reproductive, and child health and nutrition, the US Agency for International Development (USAID), United Nations Children's Fund (UNICEF), World Health Organization (WHO), World Food Programme (WFP), Global Alliance for Vaccines and Immunizations (Gavi), the Bill and Melinda Gates Foundation (BMGF), World Vision, and Mercy Corps, among others, have worked with civil society partners (such as BRAC, Grameen, and Dimagi), private partners (such as Microsoft, GSMA, and Eleven01), and governments to apply mobile phones, drones, Internet of Things (IoT) sensors, blockchain, AI, and data platforms to improve water monitoring, sanitation, medical supply deliveries, end-to-end vaccine tracing, and workload reduction for frontline health workers.

South Asia spearheaded an earlier phase of digital health, and all countries in South Asia have some discrete elements of digital health systems, including so-called eHealth and mHealth. In 2006 India triggered globally scaled adoption of open-source District Health Information Software 2 (DHIS2) in Kerala and then in more than 10 other states. The software was later adopted by some 70 countries with the support of international partners.[2] Sri Lanka, which intentionally built local technology capabilities,[3] was the first country to adapt DHIS2 for COVID-19 case detection, situation reporting, active surveillance, and response. The system is currently operational or in development in 50 countries, mainly in Africa and Asia.[4] Most South Asian governments have central health information systems for administrative efficiency, and many countries are either piloting or scaling up initiatives that connect doctors, hospitals, and patients.

That said, the new phase of digital health is qualitatively different. Across developed and developing economies, the stated objectives of governments and major global technology players in harnessing digital and converging technologies are shifting. First, they are moving from performance monitoring and delivery management by the public sector at the health system level to creation and development of private-led health innovation ecosystems. Second, there is a shift from aggregation and anonymization of health medical records for public administration purposes to use of individuals' data for risk identification, preventive services, and personalized care. And, third, there is movement away from one-way data collection from medical front lines to the facilitation of data exchanges and interoperability across public and private platforms to enable services by multiple participants in health service delivery chains, such as public and private payers, providers, and device manufacturers.

Government visions, policies, and capabilities vary widely across the countries of South Asia, as does the investment climate that fosters innovation in the private sector. Based on global comparisons of digital health, supplemented by information on governments' digital health strategies and policy frameworks as well as the broader private sector investment climate, South Asian countries can be placed along a spectrum that runs from relatively high to low preparedness for digital health:

• India, not covered by any international digital health assessment, can be considered relatively highly prepared in view of the evident digital government capabilities and the vision of the National Health Stack, which will be implemented through the planned National Digital Health Mission. However, foundational legal frameworks, such as safeguards around data protection and privacy, are still in the early stages.

- Bangladesh has a strong track record in e-health policies and practices, integrating technology into its health system to address disparities. Strong political support for this digital health strategy, especially for e-health standards and interoperable health information exchange platforms, has influenced the rapid proliferation of such initiatives in the country. A key challenge is that the availability of digital health services alone is not sufficient to overcome the underlying socioeconomic barriers to their use, such as the sophistication required to articulate health needs or to have the resources needed to afford them. Bangladesh's achievements in e-government and innovation capabilities are more modest; it has the second-weakest business environment in South Asia. The legal framework for data rights is not strong.

- Bhutan, Maldives, Nepal, and Sri Lanka have articulated aspirations on digital health, but their institutional capacities and workforce, including on health informatics, are not yet mature. Nepal has a broader cross-sectoral digital strategy led by its Ministry of Information and Communications. In Sri Lanka, although national implementation of policies is at an early stage, different sectors have spearheaded the adoption of an open-source national health information system and nurtured a local talent base. These countries, though, have little preparedness in legal frameworks as safeguards against potential risks to digital health.

- Afghanistan and Pakistan have emerging bottom-up e-health initiatives. A small health-related industry base is located in Pakistan, which also has in general a more business-friendly environment. However, it lacks a government vision, leadership, budgeted plans, and the capacities required to harness the potentials of health technology to benefit the population.

Without building the required public sector capabilities to harness the power of digital health, a largely private sector–led deployment of converging technologies in the health sector will exacerbate health inequalities in the region. Profit-seeking businesses are incentivized to serve (and extract value from data of) wealthy customers. However, policy approaches that solely protect the poor and vulnerable without harnessing private sector innovation and encouraging sustainable business models would not close these gaps either.

The health sector's response to COVID-19 has revealed the weaknesses in the region's health infrastructure and services, which are severely underfunded, especially those in the public sector. The health care system has been overwhelmed, with shortages of staff as well as medical and personal protective equipment. Technology has been leveraged for effective communication and, to a certain extent, for teleconsultation and contact tracing. Pakistan has deployed technology to identify pockets of infection to initiate "smart lockdowns." India has been able to mobilize its technological strengths, especially in vaccine research, and has witnessed a wave of innovation by local start-ups and industries, ranging from low-cost portable ventilators and sanitizers to telemedicine platforms and assistant robots. Yet India's testing capacity remains suboptimal, the secondary health burdens are mounting, and its overall weak health infrastructure is severely strained.

Overall, however, technology-enabled responses to COVID-19 to protect the population have proved effective when anchored in local and indigenous capacities cultivated over the long run, as in some Indian states. With relatively high levels of public funding, strong public health infrastructure, granular local information on vulnerable groups, and effective leadership (which drew lessons from an earlier lethal infectious disease outbreak and flooding), the state of Kerala in India was able to respond effectively at the outset of the pandemic.

EDUCATION

South Asia's public education systems generally bypassed the early stage of education technology, such as introducing information and communications technology (ICT) for teachers and students by deploying computers, e-learning, and learning management systems in classrooms. Instead, computer labs were introduced in selected secondary schools, with limited integration of ICT skills into the core curriculum. The lack of connectivity, infrastructure, equipment, and electricity in most public schools hampered the extensive use of technology tools. A notable exception was the state of Kerala in India, which began the introduction of ICT in a systematic way in public school education over two decades ago and gradually rolled out the initiative from high schools down to primary schools, upgrading it as technology changed. Kerala now has a platform built fully on free open-source software, with teacher-developed content integrated into the curriculum and lesson plans, extensive teacher training and ongoing technical support, as well as the provision of infrastructure in high schools.

The higher education systems in South Asian countries are served by national research and education networks (NRENs), which are well developed in the region. These networks provide higher education institutions with broadband. And in addition to connectivity, they also provide digital skills training for faculty and students to transition to online and blended learning, a design for campus networks, and technical support. There is wide variation in the capacity and reach of the NRENs; India's National Knowledge Network is the most advanced. A state-of-the-art multigigabit network that connects education, research, and government facilities across India down to the district level with reliable and secure links, it is designed to be dynamic and scalable. Plans are now under way for an upgrade to a speed of 100 gigabits per second. NRENs in Bangladesh, Pakistan, and Sri Lanka also have significant capabilities, whereas those in Afghanistan and Nepal are still nascent.

The current phase of education technology, both globally and in the region, is qualitatively different from the earlier phase of stand-alone technology tools (such as computers, e-books, and smartboards) and is driven by digitization and connectivity as well as the explosion of knowledge resources (these resources are, however, mostly still in the English language). There is also a shift from products to services. Technology tools are aimed at specific user groups such as teachers, students, administrators, and employers (for skills training), and at different functions such as teaching-learning tools and resources, learning management systems, and assessments. A variety of business models have evolved, including many subscription-based services with prepackaged content, where often the collected user data are the real value to the company because they can be monetized. The range of technologies has also expanded and includes AI-based learning support for greater customization and "personalized" learning experiences, mixed reality-based learning, game-based learning, and mobile learning. At the global level, this convergence of education platforms and individualized assessment and learning profiles is paving the way for the expanded use of cognitive computing applications in education and learning. Relevant use cases include their deployment in (1) smart administration to increase efficiency; (2) smart curriculum for developing digital content courses; (3) smart learning systems for changing the way in which learning systems offer a variety of options, depending on the interests or needs of the learners; and (4) smart learner support, helping to analyze the learning behaviors or problems of learners, thereby helping instructors modify content and direction.

The South Asia region has seen enormous growth in private education technology companies. They are mainly business-to-consumer (B2C) firms, in which the consumer consists of schools, mainly private, and students or their parents. India is emerging as a leader in education technology together with global technology companies and e-commerce platforms in China, the United Kingdom, and the United States. Nascent private edtech companies have opened in Bangladesh and Pakistan. Indian edtech companies now number almost 4,500, the fourth largest constellation of such firms in the world (Inc42 2019). Bangladesh, Pakistan, and Sri Lanka have a much smaller number of companies. Nepal has about 28 education start-ups.[5]

However, the adoption of technology is still driven largely by private spending by more affluent households in urban areas who send their children to elite private schools. Products largely focus on private tutoring and test preparation, as in China and other Asian countries. A significant share of this spending is for mobile phone apps for after-school use by students, usually involving online tutoring or test preparation. Although "personalization" occurs to a certain extent in that learning support is tailored to the student's level, overall learning goals are conditioned by the structure of examinations (for which these tools provide support), which still focus on rote learning and memorization, as is common in other Asian markets, including China, Indonesia, and the Republic of Korea.[6] Another feature of the Chinese edtech market, which may have parallels in South Asia because of the similar priority given to formal education, is the heavy investment by educated urban parents in early childhood learning tools and games. Early childhood learning plays a critical role in future educational success because children are better prepared to enter school. Most learning resources are in English to suit the clientele. Nevertheless, expansion into vernacular language education resources and into rural areas, especially for homework and test preparation support for children from less affluent backgrounds, is possible.

Private sector investment trends in the region and globally indicate that a new growth area is the reskilling or upskilling of workers in view of the anticipated disruptions of industries by new technologies. The current education system is overdue for a transformation to adapt to the needs of the Fourth Industrial Revolution with its fast-changing nature of jobs and short shelf life of skills. Technology platforms can reduce skill mismatches by better anticipating and preparing for emerging skills and jobs and by developing real-time labor market information systems using AI and big data analytics to help job seekers match their skills with those employers are seeking. Training institutions will be needed to focus on the regular renewal of curricula to match emerging industry needs to support lifelong learning of the existing workforce. At the same time, national qualification systems, learning pathways, and recognition and validation systems of learning should also be upgraded to motivate learners and workers to pursue ongoing learning.

The technologies attracting private sector investment are AI-based learning and personalized learning; mixed reality and immersive learning, both augmented reality and virtual reality; and game-based learning. Although translation technologies are being developed and have a potentially huge market in the education sector in South Asia, they are not yet deployable at scale in a cost-effective manner. The use of English as the language of instruction in private education institutions also reduces the demand for translation tools in the most profitable segments of the market.

Digital public platforms in the education sector are being developed in almost all countries in South Asia. India has built a sophisticated technology platform using the India Education Stack, which has a layered architecture: from the basic infrastructure layer to the knowledge, content, and measurement layers, leading eventually to personalized education to meet different purposes (teacher training, early learning through higher education plus skilling, leadership education, and capacity building).[7]

In India, the National Digital Infrastructure for Knowledge Sharing (DIKSHA) for school education is now operational on the India Education Stack.[8] Heralded as having great potential to transform the use of technology in teaching as well as teacher training, DIKSHA is expected to include in-class resources, teacher training materials, assessments, and communication tools. QR (Quick Response) codes linked to textbooks enable teachers, in principle, to show students resources linked to a particular lesson and available on the platform.[9]

A critical concern here is that mostly private sector firms will take advantage of these platforms and aim them at students from higher-income families who can afford the technologies or at those seeking professional development in the formal sector. The challenge will be to use these platforms to develop educational programs that will improve the education of children in public schools as well as for children who are out of school and adults lacking functional literacy.

The mere creation of digital platforms for the education sector is not enough, however. Experience in India and other countries has shown that several years of ongoing support for teachers and continual teacher involvement in the preparation of materials are needed to integrate the resources available on these platforms into classroom teaching to achieve improvements in learning. In short, the starting point must be the needs of the primary users—the teachers and students. The complementary investments in their skills, knowledge, accessibility to devices, and connectivity are necessary to use digital platforms effectively.

The COVID-19 pandemic has highlighted many of these inequalities in the use of technology. Most school systems could not use digital public platforms and had to switch to television, radio, YouTube, and WhatsApp messages. In Pakistan, television lessons were combined with SMS (short message service) texts to students to help them improve their learning. The state of Kerala used infrastructure and content already available to deliver learning through multiple channels, such as its special internet-based education television channel and YouTube. Higher education NRENs were able to maintain access to their portals.

Overall, however, during the pandemic digital technologies could not be deployed at scale in the public sector, especially in school education, because of the lack of digital content for each class and subject; the limited training of teachers to use this content effectively; the limited digital skills of students to access and use digital information; the scarce access of students to affordable broadband and digital devices, including in many cases television or radio, as well as to a regular electricity supply; and an inadequate capacity to manage digital systems in education. For example, even in Kerala content could at most be delivered for 1–2 hours a day, with teachers expected to follow up with students later. A 2020 survey of northern and eastern Indian states showed that 80 percent of students studying in government schools had received no education at all since March 2020, with most not even having received textbooks (Oxfam 2020). Even in higher education, a survey conducted in Sri Lanka by the Asian Development Bank revealed that less than half of the student respondents had laptops; 50 percent said that data packages were not affordable or only somewhat affordable; and 70 percent said that poor internet connections limited their access to online classes (Hayashi et al. 2020).

By contrast, deployment of technology in the private education sector took off rapidly during the COVID-19 pandemic, with an enormous increase in the use of private applications and learning tools by students at home. Meanwhile, private schools and universities serving more affluent households were able to shift to online learning.

Another important issue is related to the collection, storage, use, and transfer of data collected through education platforms from potentially tens of millions of children, young people, and teachers. These data can include sensitive personal information, including possibly about the families of students. The current regulatory regime for personal data protection is weak in South Asia (see chapter 5). Mapping the data sources, paths for data flows, linkages between different actors on the platform, and use of the data, and establishing transparency are critical to ensure safety, accountability, and trust.

SOCIAL PROTECTION

Two key technology parameters underpin social registries: (1) the information processes and systems that enable a harmonized process for outreach, intake, registration, and eligibility determination of potential beneficiaries for one or more social programs and (2) the foundational IDs that serve to deduplicate the identity of applicants and to facilitate interoperability with other information systems.

Application to social registries and delivery of benefits or services to eligible beneficiaries can profit from technologies such as web-based applications, automated cross-verification with other data systems, a government-to-person (G2P) payment processing system, and G2P monitoring and communication. These elements are at various stages of development in South Asia. Their relative strengths were on display during the COVID-19 pandemic as governments had to quickly ramp up cash transfers to offset income losses for a large segment of the population. Because of the crucial importance of social registries and efficient payment systems for inclusive and dynamic social protection in the face of future challenges, their performance needs to be strengthened in all countries.

Foundational IDs are in place in Bangladesh, Bhutan, India, Nepal, Pakistan, and Sri Lanka, but population coverage varies greatly. Bangladesh plans to establish digital identity from birth, using its vaccination and birth registration processes, and is exploring infant biometric technologies. The process of expanding coverage of national ID and modernization of civil registration is under way in Nepal. Bhutan has provided every adult with a unique citizen identity number and is now embarking on the National Digital Identity initiative under its Digital Drukyul Flagship. Nevertheless, concerns exist in all countries about data privacy, data security, speed of implementation and expansion, and use of data.

The use of digital technologies for payments is, however, severely constrained by the relatively low levels of financial inclusion in many countries. About 80 percent of adults in India had a financial account in 2017, with a relatively small gender gap (figure 3.3, panel c). However, less than half had made digital payments in the previous year (panel b), and almost half of the accounts were inactive during the previous year (panel a). The proportion of inactive accounts was high in Sri Lanka as well, at about one-third. About 50 percent of adults in Bangladesh had an account, but the gender gap was significant; 45 percent held accounts in Nepal.[10] In the last few years, payments to civil servants and large social protection programs such as the social security allowances and the prime minister's employment program have begun shifting from cash delivery to bank payments. The lowest shares of account holders using digital payments are observed in Pakistan (21 percent), again with a large gender gap, and in Afghanistan (15 percent, not shown in figure 3.3).

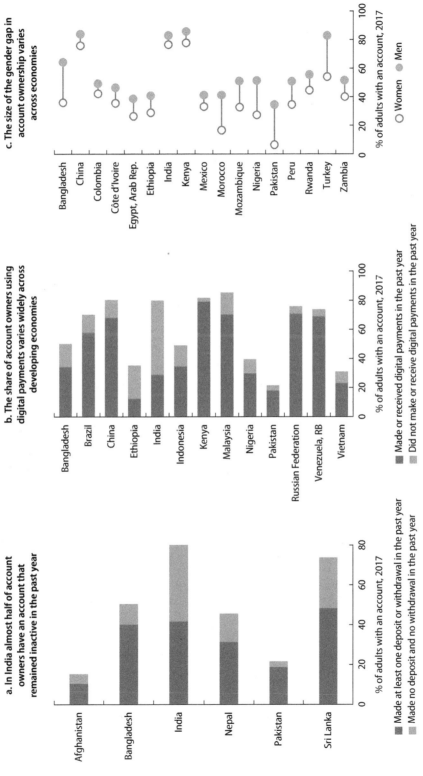

FIGURE 3.3 **Account Ownership and Digital Payments: Selected Countries, South Asia**

a. In India almost half of account owners have an account that remained inactive in the past year

b. The share of account owners using digital payments varies widely across developing economies

c. The size of the gender gap in account ownership varies across economies

■ Made at least one deposit or withdrawal in the past year
■ Made no deposit and no withdrawal in the past year

■ Made or received digital payments in the past year
■ Did not make or receive digital payments in the past year

○ Women ● Men

Source: World Bank, Global Findex (database), 2018, https://globalfindex.worldbank.org/sites/globalfindex/files/countrybook/Nepal.pdf.

Pakistan's National Social and Economic Registry (NSER) covers almost 90 percent of the population (Leite et al. 2017). Information is collected by competitively contracted firms allocated to specific geographic areas, with a separately contracted firm to supervise the fieldwork. The database is used by multiple programs at the federal and provincial levels that make enrollment decisions based on program-specific eligibility criteria and budgetary space. Although based on census "sweeps" every few years rather than offering open, ongoing registration, the exclusion of newly impoverished households is limited because of the high level of coverage.

The integration of NSER with the National Database Registration Authority (NADRA), which issues the identity card, has made identification of beneficiaries relatively easy (and avoids duplication of payments), as well as the exclusion of those ineligible for payments. The state-of-the-art registration as well as the payment system using biometric verification are key strengths. Payments are conducted through two competitively selected commercial banks with retail agents in small shops. During the COVID-19 pandemic, the government was able to provide one-time financial assistance to about 12 million families (approximately 80 million people). Families could send an SMS to the special hotline together with their ID number, and those meeting predetermined eligibility criteria, established using data analytics, received a confirming SMS. Inclusion criteria were expanded to those who might be more vulnerable during the lockdown. Operationalizing the payments in the context of COVID-19 lockdowns required careful coordination across different levels of government and financial institutions, as well as communications to explain the process to recipients.

India's combination of JanDhan (bank accounts for receiving direct benefits), Aadhaar (unique identification), and a mobile phone number—the so-called JAM trinity—also played a crucial role in providing social assistance to the tens of millions of people who lost their jobs in urban areas. However, because over 800 government schemes at the central and state levels and many different ministries were involved, the ability to respond in a focused, rapid manner was much more difficult during the COVID-19 pandemic. A national social registry with coordination across programs does not yet exist. With high levels of internal migration, the identification of beneficiaries by state of origin (such as for the public food distribution system) tended to exclude migrants. Approximately 35 percent of migrant workers did not have access to a bank account in 2017–18, even after the rollout of the JanDhan accounts, according to a survey conducted by the Centre for Digital Financial Inclusion (CDFI 2020).

Sri Lanka was able to leverage its national social registry to augment payments to the poor. In Bangladesh, there is no national social registry, thereby limiting the ability to identify poor or vulnerable households and deliver cash quickly. The country is, however, moving toward establishing a national social registry using the database from the national identity system and ensuring interoperability across platforms delivering social programs.

For the future, the move toward dynamic social registries that would enable continuous enrollment of households, depending on their changing status, should be a priority for all countries. Strengthening the G2P systems in partnership with the private sector is a second priority. Furthermore, the use of data analytics and predictive models, relying on multiple different technologies, including satellite imagery and geospatial technology, could help to identify clusters of vulnerability and improve the targeting of social assistance programs.

CONVERGING TECHNOLOGY CASE STUDIES IN SOUTH ASIA

Two case studies undertaken in the context of this study—identification of stunting among young children and transforming learning into a more adaptive and personalized service—highlight the immense potential for using these technologies to address endemic human capital challenges in South Asia (Pauwels 2020). However, they also highlight the significant risks discussed later in this chapter.

Rates of stunting among young children in South Asian countries are among the worst in the world, next only to those in Sub-Saharan Africa. Microsoft and Welthungerhilfe, a German private aid organization, are developing a mobile application on a smartphone equipped with a set of infrared sensors to test children in three states in India for stunting using facial and body recognition technology (Microsoft News 2019). The smartphone captures 3-D measurements of the child's height, body volume, and weight ratio, head and upper arm circumference, as well as facial features. The application also collects personal data on each child, including family name, guardian, age, sex, birth date, and location. When connectivity is available, the information can be sent immediately to remote nutritionists and digital specialists who can evaluate the scans using an algorithmic platform to assess the health and nutritional status of the child. If mobile connectivity is not immediately available, the information can be uploaded later. Data, including facial and biometric features, are stored on Microsoft's cloud platform, Azure.

The potential benefits of this application are greater accuracy in collecting the information, automation of the process, and the possibility of releasing health fieldworkers and funding for other tasks. However, the risks include an inaccurate diagnosis if the datasets over which the AI algorithm was trained were not large enough to be relevant to the specific target population. Other concerns are about data security, privacy, and consent. And yet another concern is about technological ownership and benefit sharing if the private company holding the technology does not transfer the technology, skills, and data to local doctors and hospitals. Finally, although useful, the phone-based remote diagnosis is just the first part of process. It must be followed up with concrete actions to address the problem.

Converging technologies and AI offer the possibility of transforming learning into a more adaptive and personalized process. In addition to just adjusting to the child's learning levels (based on responses to quizzes and tests), the use of affective computing allows algorithms to sense how children are reacting to learning content based on monitoring their faces and behavior and to vary the form of instruction to fit their needs.

Such an advance could help tailor education to each student and increase the efficiency of the educational system. The process is comparable to that of a "classroom" situation in which the teacher can respond to the noncognitive responses of the child (emotional, motivational, and so on) and adjust teaching accordingly.

However, there are multiple risks. The system reduces complex emotional states to a basic coding of facial movements and makes inferences from that about learning. Studies have found that efforts to read a person's internal states from facial movements alone are at best incomplete or at worst may entirely lack validity. The scientific validity of the Facial Action Coding System being used by education technology companies in affective computing is also being questioned.

On the pedagogical side, there is the question of what the children are being trained for and what effect this has on overall child development. The constant monitoring may have a chilling effect on children when they realize they are being "scored" by machines, and it may affect peer relations and reinforce biases and prejudices. Furthermore, it is not clear that AI-enabled applications are better at developing creativity and resilience compared with peer-to-peer interaction and human mentorship.

In addition, there are issues about how the extensive data collected about students and their families, including payment history, are stored and used by the provider of the service. Questions such as who owns the data and whether it can be used for commercial purposes by the company or sold to others, including for surveillance purposes, are paramount. Companies expand their facial biometrics collection through the provision of such services, which reinforces their position in the market and contributes to their "legitimacy."[11] There is also the issue of potential explicit or implicit biases in the algorithms and whether they will unfairly typecast a student based on incomplete or biased assessments.

Data-Driven Decision-Making in the Human Development Sectors

Health, education, and social protection pose great challenges for data collection through traditional means and even more so for data analysis and use in decision-making. These challenges stem from the large geographical spread of service locations, poorly paid and relatively unskilled frontline providers, and weak monitoring. Ubiquitous and faster digital connectivity, coupled with data collected "automatically" from devices and sensors on human beings and inanimate objects, has transformed this situation. These trends, combined with satellite data and geospatial information becoming increasingly available, are enhancing the scope for data and evidence to support real-time decision-making and monitoring and evaluation of development interventions in the social sectors.

Applications range from the individual organization level (such as hospitals or universities) to entire government departments. For example, analysis of student performance data can enable university faculty to identify students at risk of dropping out and suggest supportive actions. Data-driven decision-making techniques can also target individual functions such as supply chain management, logistics, and monitoring student and teacher attendance, as well as systemic issues such as disaster response for the entire health care and education services or planning the resilience of the service to future shocks. Allocation of health provider staff can be matched to the arrival times of patients. Rapid feedback from beneficiaries and users, using text and photos, can alert administrators to deficiencies in service delivery, equipment breakdowns, and supply shortages.

Geospatial data can be particularly effective when integrated with planning, census, and survey data for poverty mapping and decisions on allocations of facilities and resources, including for services such as telemedicine and tele-education. The potential of combining granular geospatial data with socioeconomic data to microtarget some of the entrenched human capital challenges in South Asia, such as stunting and adult illiteracy, is enormous. Geospatial technologies have proved to be especially effective tools in postdisaster relief and are increasingly prominent in the sectors that contribute to human capital such as agriculture, water, energy, and transport. The combination with other technologies, such as AI and big data, as well as advances in miniaturizing sensors, cloud technology, and high-speed computing, are pushing the frontiers of geospatial technologies.

Data-driven decision-making can therefore vastly improve the effectiveness and responsiveness of services to build human capital. As with other aspects of technology, however, success depends not on the specific technology tool. The most critical factors are the mindset of government leaders and their willingness to use the potential of digital tools to redesign citizen-centric services; the technical skills of administrators to absorb and use technologies; and the policy frameworks and their translation into governance, legal, and regulatory mechanisms. The infrastructure for data, comprising both the hardware and the rules and institutions that govern the sharing of data in a safe and secure manner, is critical to making use of the full potential for data-driven decision-making. Currently, all these aspects are weak in South Asian countries, creating an opportunity for engagement to strengthen the capacity of governments.

Conclusions

This technology landscaping exercise has revealed that the use of converging technologies in health and education is advancing in many South Asian countries, with India in the lead, especially in the private sector. The promise of converging technologies is greatest in the health sector. The first wave of the COVID-19 pandemic demonstrated a rapid shift toward digital means for delivering services, with successes in rapid digital payments for scaled-up social assistance programs.

However, the existing trends also indicate that the deployment of these technologies is likely to exacerbate the existing inequalities in human capital unless countervailing public policies are put in place and governments address the constraints facing marginalized and poor populations when trying to access services through digital means. The private sector is driving the development of these technologies, primarily motivated by users among the wealthiest households.

Although digital public platforms (or public-private platforms) provide one way of leveraging digital technologies for the poor, the barriers to use must be explicitly addressed from the beginning. Among these barriers are first-mile access to digital infrastructure by schools and health centers in poor communities, as well as by women within households, and the availability of local content (especially important in education but also for delivery of health services). Thus design of these platforms must explicitly take into account the conditions and skills of the end-users.

Integrated social registries are a powerful way to support the equity agenda in South Asia and build long-term resilience as well as a government's capacity to use digital platforms to improve service delivery. The COVID-19 experience demonstrated the ability of several South Asian countries to deliver social assistance on a massive scale, building on existing infrastructure and programs.

The use of converging technologies, underpinned by AI-enabled data collection and analysis and data-driven decision-making, requires building a strong capacity in the public sector (leadership, vision, and skills), as well as creating the technology and data governance systems and processes that will protect individuals and groups. However, AI-enabled converging technologies could harm people by targeting special groups, undertaking exfiltration of data for other purposes, and posing cybersecurity threats. Their use among children and populations with limited literacy or awareness of the technology must be regulated. A citizen-oriented technology governance system and a regulatory framework and architecture for the use of data are therefore essential to ensuring that the potential for improved service delivery is not misused to reduce voice, agency, and empowerment.

Notes

1. Section 135 of India's Companies Act 2013 stipulates that companies in India having a net worth of Rs 5 billion (equivalent to around US$69 million) or more during any financial year must formulate a CSR policy and spend at least 2 percent of the average net profit of the company over the three immediately preceding financial years on CSR activities.

2. Including Norad, BMGF, The Global Fund, US Centers for Disease Control (CDC), Gavi, UNICEF, WHO, and the U.S. President's Emergency Plan for AIDS Relief (US PEPFAR). See the DHIS2 Fact Sheet, https://s3-eu-west-1.amazonaws.com/content.dhis2.org/general/dhis-factsheet.pdf.

3. Through higher education, provincial pilots for maternal and child health, and integration of a geographic information system (GIS) and adoption in 2012. See Manoj et al. (2013).

4. See the DHIS2 COVID-19 Surveillance, Response and Vaccine Delivery Toolkit (https://dhis2
.org/covid-19/). Open-source communities in collaboration with DHIS2 are also working on
interoperable health information exchange architecture. See the OpenHIE COVID-19 Task
Force, https://wiki.ohie.org/display/SUB/COVID-19+Task+Force.

5. Tracxn, https://tracxn.com/explore/EdTech-Startups-in-Nepal.

6. In China, the highly competitive university examinations "are the most influential driver of
China's US$50 billion after school tutoring and test preparation industry. China's two largest
education companies both began as offline tutoring centers, eventually offering an online
component" (Omidyar Network 2019).

7. The system was built using the Sunbird digital infrastructure, an open-source, configurable,
extendable, modular learning management infrastructure designed for scale and available
under license from the Massachusetts Institute of Technology (http://Sunbird.org). The
Education Stack is currently maintained by the EkStep Foundation in partnership with India's
Ministry of Education. The terms of the agreement are not publicly available.

8. DIKSHA is maintained by the EkStep Foundation.

9. The installation of QR codes by itself does not represent advanced instructional technology
because it can lead to passive teaching-learning.

10. Work Bank, Global Findex, https://globalfindex.worldbank.org/sites/globalfindex/files
/countrybook/Nepal.pdf; ADB (2020).

11. Affectiva, a leading facial emotion recognition company that is entering the education field,
already had a database of more than 8 million faces in 2019. Intel and Microsoft Azure are also
in the field of facial coding that could be used for education. Microsoft Azure is the platform
used for storing the facial and biometrics features of Indian children in order to diagnose
stunting, indicating how datasets from different products and services could be merged without the knowledge of users to develop new products and services and for broader goals such
as "social scoring" and "conditioning behavior" for commercial or political purposes.

References

ADB (Asian Development Bank). 2020. *Nepal: Macroeconomic Update.* Vol. 8, no. 2, September.
Mandalyuong, Phillipines: ADB. https://www.adb.org/sites/default/files/institutional-document
/637301/nepal-macroeconomic-update-202009.pdf.

CDFI (Centre for Digital Financial Conclusion). 2020. *How Migrants Send Money Home? Study
of Profiles, Preferences and Potential.* http://www.cdfi.in/assets/images/CDFI_Migrant
_Remittance_Study_Report.pdf.

Deloitte and Healthcare Federation of India. 2016. *Medical Devices: Making in India—A Leap
for Indian Healthcare.* https://www2.deloitte.com/content/dam/Deloitte/in/Documents/life
-sciences-health-care/in-lshc-medical-devices-making-in-india-noexp.pdf.

Hayashi, R., M. Garcia, A. Maddawin, and K. Hewagamage. 2020. "Online Learning in Sri Lanka's
Higher Education Institutions during the COVID-19 Pandemic." ADB Briefs, No. 151, September,
Asian Development Bank. https://www.adb.org/sites/default/files/publication/635911/online
-learning-sri-lanka-during-covid-19.pdf.

Inc42. 2019. "What Will Be the Biggest Edtech Trends in 2020?" https://inc42.com/features /what-will-be-the-biggest-edtech-trends-in-2020/.

Leite, Phillippe, Tina George, Changqing Sun, Theresa Jones, and Kathy Lindert. 2017. *"Social Registries for Social Assistance and Beyond: A Guidance Note and Assessment Tool."* Social Protection and Labor Discussion Paper No. 1704, World Bank, Washington, DC. https:// openknowledge.worldbank.org/handle/10986/28284.

LIRNEAsia. 2019. *AfterAccess: ICT Access and Use in Asia and the Global South.* https://lirneasia .net/2019/05/afteraccess-asfindeia-report3/.

Manoj, S., et al. 2013. "Implementation of District Health Information Software 2 (DHIS2) in Sri Lanka." *Sri Lanka Journal of Bio-Medical Informatics* 3 (4): 109–14.

Microsoft News. 2019. "Germany's Welthungerhilfe Rolls Out Pilot Project in India to Tackle Malnutrition with Microsoft AI." https://news.microsoft.com/en-in/welthungerhilfe-child -growth-monitor-pilot-project-india-malnutrition-microsoft-ai/.

Omidyar Network. 2019. "Scaling Access and Impact: Realizing the Power of EdTech: China Country Report." March. https://omidyar.com/scaling-access-impact-realizing-the-power -of-edtech-2/.

Oxfam. 2020. *Status Report—Government and Private Schools during COVID-19.* https://www .oxfamindia.org/sites/default/files/2020-09/Status%20report%20Government%20and%20private %20schools%20during%20COVID%20-%2019.pdf.

Pauwels, E. 2020. "Converging Technologies for Human Capital." Paper prepared for South Asia Technology for Human Capital Study, World Bank, Washington, DC.

World Bank. 2020. "DE4A: Digital Economy for Africa Initiative: Digital Economy for Africa Country Diagnostic Tool and Guidelines for Task Teams, Version 2.0." World Bank, Washington, DC. https://pubdocs.worldbank.org/en/694441594319396632/DE4A-Diagnostic-Tool-V2 -FINAL-JUNE-24.pdf.

World Bank. Forthcoming. *Digital Economy Assessment for South Asia.* Washington, DC: World Bank.

CHAPTER 4

Deploying and Utilizing Human Capital: Implications of the Converging Technology Revolution for Employment and Innovation

Introduction

Converging technologies will have a great impact on the preparation and protection of human capital, but they will have an equally if not more important role in the utilization of human capital in the job market. The potential of the Fourth Industrial Revolution in production to fundamentally alter the demand for labor is immense. On the one hand, it will disrupt and displace jobs. On the other, it will possibly create higher-quality jobs. Without adequate policy measures in place for massive reskilling and for dynamic social protection, the impacts on human capital could be devastating.

Meanwhile, advanced, highly specialized human capital contributes to the development and deployment of new technologies across the economy, as part of the innovation ecosystem. The production, sharing, and utilization of this scarce human capital is essential if countries are to benefit from the ongoing technological revolution. The relationship is rendered even more complex by the fact that converging technologies are changing the way in which innovation is taking place globally.

Impact of New Technologies on Labor Demand in South Asian Countries

The deployment of technologies in production may replace or complement human labor, but many other factors also affect overall labor demand. Morover, technology adoption is itself driven by a large number of factors, and the impacts on different countries will depend on their sectoral composition, the relative cost of labor to capital, and the extent of competition from other countries.

To accelerate development, emerging economies need to leapfrog industrialization to the high-tech economy (UNCTAD 2021). Leapfrogging will require prioritizing investments in people for the acquisition of higher-level skills, but it also will, in turn, create more demand for jobs in the local service economy. In South Asian countries with large shares of youth, there is heightened urgency to provide youth and migrant workers from disadvantaged groups with a mix of hard and soft employability skills so they can reap the benefits of demographic dividends. To combine human intelligence to innovate with the computing power of machines, the workforce would have to acquire twenty-first-century skills such as critical thinking, creativity, communication, collaboration, problem-solving, cross-cultural literacy, work ethic, empathy, and socioemotional and digital intelligence.

COVID-19 has exposed the fragility of the world's supply chains for medicines and medical products, food, energy, vehicles, telecom equipment, electronics, and countless other goods and services. In response, certain companies, notably early adopters of Fourth Industrial Revolution technologies, have begun to reconfigure their sourcing and manufacturing footprints for greater reliability and resilience. Others are accelerating the adoption of digital work instructions, augmented reality-based operator assistance, and simple, inexpensive retrofit automation. This reorganization of production patterns, which is being accompanied by increasing servicification of supply chain networks, can either increase or decrease the number of jobs. For the South Asia region, developing globally competitive manufacturing hubs is one of the biggest opportunities to operate in international markets, double its manufacturing gross domestic product (GDP), create new high-value service jobs, and provide long-term employment and skill pathways for millions this decade.

How can India, in particular, and the subregion, as a whole, take advantage of these shifts? Mature value chains (such as pharmaceuticals, automotive components, fast-selling technology products, and software) and service sector jobs (such as health care, tourism, financial services, logistics and supply chain coordination, information technology outsourcing, and creative industries), both of which rely on sophisticated capabilities and healthy supplier ecosystems to serve domestic and export markets, must scale up. A second group of value chains, which mainly produce for domestic markets (such as food processing) but lack scale, productivity, and technological sophistication, must transform to compete. Finally, entry into emerging value chains (such as the "sunrise" sectors in energy storage, electric vehicles, and the bioeconomy) will require strategic partnerships with global consortia to access the technology and capital needed to establish local manufacturing capacity. Although the unique circumstances of the pandemic have elevated resilience, agility, and flexibility in operations as top strategic priorities, the medium-term agenda for the South Asia region will be to lift the productivity of manufacturing value chains to global standards.[1]

Meeting the growth in the domestic and global demand for products tends to increase the demand for jobs. In a recent study, the Asian Development Bank decomposed these factors and found that between 2005 and 2015 the introduction of new technologies displaced 60 percent of jobs in Bangladesh and 40 percent of jobs in India (the only two countries covered in the study) (ADB 2018). However, the demand for new jobs increased by 80 percent, primarily because of the increase in domestic demand (figure 4.1).

Looking forward, will the ongoing technological revolution have an even more profound impact on job displacement, including net task relocation away from South Asia? And can post-COVID-19 growth in aggregate demand offset not only temporary but also structural job losses? A particularly worrying aspect of the COVID-19 pandemic is that young people, trainees, and fresh graduates are disproportionately hit by joblessness because many companies are retaining experienced employees and have cut back on hiring.

Estimates of potential job losses and displacement are relatively scarce and variable for the South Asia region. Several estimates for India indicate how the adoption of technology and its impact on jobs can vary substantially across different industries. The World Bank's *World Development Report 2016: Digital Dividends* estimated that 69 percent of jobs in India were at risk of being automated, although this rate dropped to 44 percent when adjusted for potential adoption speed (World Bank 2016). Other analyses carried out by the International Labour Organization (ILO) and the Organisation for Economic Co-operation and Development (OECD) for India suggest that capital-intensive manufacturing industries (such as the automobile industry) are more likely to adopt advanced automation and robotics (ILO 2018; OECD 2020). Although this industry had the highest employment growth rate over the last decade, it still employs only a relatively small share of the labor force. In the informal sector, where over 80 percent of the labor force is employed, the adoption of advanced technologies is impeded by lack of financial capital and necessary skills. The agriculture sector, which is the largest employer in India, has high automation potential (using the Internet of Things, digital platforms, and data analytics) to improve productivity, but low growth in the sector and the shrinking size of land holdings make widespread technology adoption unlikely. Labor-intensive industries (such as textile, leather, and footwear) are unlikely to be rapid or large-scale adopters of Fourth Industrial Revolution technologies because such technologies require dexterity, labor cost advantages, and high capital investment.[2] A potential risk for jobs is that major brands in the ready-wear garment industry may relocate their manufacturing hubs to speed up product-to-market cycles, facilitated by developments in automation, a trend that has picked up momentum during the pandemic.

Although it is unlikely that large-scale adoption of technology will occur in the informal, unorganized sector where a majority of women and the poor are employed, this sector could both benefit from and be negatively affected by diverging trends in the

FIGURE 4.1 How Technology and Other Channels Affect Jobs

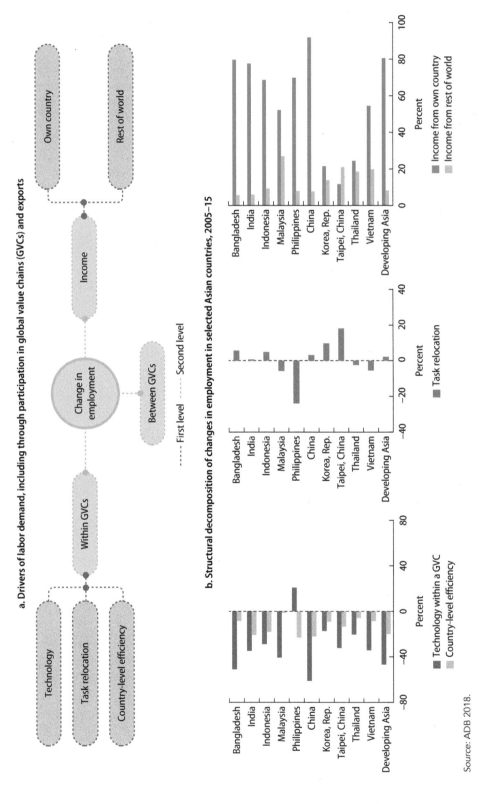

a. Drivers of labor demand, including through participation in global value chains (GVCs) and exports

b. Structural decomposition of changes in employment in selected Asian countries, 2005–15

Source: ADB 2018.

deployment of technologies. Digital technologies and new business models, such as platforms and e-commerce, allow self-employed and small producers, including food delivery, repair services, and passenger transportation, to trade goods and offer services that can help them grow and formalize, create better jobs, and raise incomes. To the extent that digitalization of government services helps to reduce the transaction costs of accessing these services, it may also incentivize formalization. On the other hand, automation within services could create new opportunities for data-driven business backed by large technology firms and potentially displace numerous small and medium enterprises.

Further analysis of the potential impacts of technology on these sectors is important because they employ the majority of the poor, who already face many hurdles in accessing better-quality jobs in the formal sector. Greater adoption of technologies requiring higher levels of cognitive, technical, and socioemotional skills could further deepen inequalities in the labor market, unless countervailing public policy measures are put in place. These include improving opportunities for better-quality preemployment education and skills training, massive reskilling of the existing workforce, as well as social protection for workers who are displaced from their jobs.

Another feature of South Asia is the low labor force participation rate of women. This low rate means that no matter how much human capital is accumulated by girls in their early and adolescent years, much of it remains underutilized and does not benefit either the country's growth or the empowerment of women. How digital technologies will affect women's labor force participation and empowerment depends on several factors. On the one hand, by enabling women to work from home, such technologies may help them to raise their income. On the other hand, lower levels of digital skills and limited access to devices and broadband can accentuate disparities between men and women, while confining women to the home may prevent broader social participation.

Although more analysis is essential to understanding the implications of the deployment of technologies in each South Asian country, especially their effects on equity, gender, and inclusion, it is also clear that South Asia needs to invest heavily in dynamic social protection systems and in adaptive reskilling of the population. It is impossible to predict exactly the future course of the economy and the sectoral reallocation of labor. But having in place the institutional mechanisms for protecting vulnerable populations is the best insurance against the backdrop of an uncertain future.

The Digitization of Innovation and the Role of Advanced Human Capital

The national innovation system plays a crucial role in the application and deployment of technologies in the economy and society. It comprises many actors and has become increasingly complex. A recent conceptualization highlights the interaction of different components of the technical infrastructure, including information and communications technology (ICT); the interplay of government, firms and entrepreneurs, research institutes and universities, and civil society and consumers; the role of finance and enabling regulations and policies; and the importance of human capital and knowledge and innovation networks (UNCTAD 2018). However, even this conceptualization does not adequately reflect the fact that the innovation system has become increasingly global as a result of international research programs and the globalization of knowledge and its diffusion through multinational companies, foreign direct investment, trade, the internet, and the international movement of skilled individuals. Furthermore, although innovation was driven earlier primarily by military and economic objectives, there is a growing realization that inclusiveness and environmental sustainability should be important goals of innovation systems. Taken together, these trends raise the importance of rethinking the concept of a national innovation system.

Apart from India, South Asian countries have nascent or nonexistent innovation ecosystems. Among other things, they lack the advanced, specialized human capital to adapt existing technologies to national and local needs, as well as the firms that can use them. One indicator of the contribution to global innovation is expenditure on research and development (R&D). India accounts for just 4 percent of the global total (compared with China which accounts for over 23 percent), while Pakistan and Bangladesh account for 0.3 percent and 0.2 percent, respectively. However, India is gaining ground rapidly in terms of new technology start-ups, including in education and health (R&D World 2020). The number of technology unicorns (privately held start-up companies with a value of over US$1 billion) in India for the period 2009–18 was 14. This figure puts India in fourth place globally, after the United States (261), China (114), and the United Kingdom (21)—see World Bank (2018). India therefore has strong potential if it can improve its high-level technical human capital and innovation capability, including the overall environment for innovation.

South Asian countries lag in the accumulation of advanced human capital as shown by the slow growth in the overall tertiary education ratio, advanced degrees in natural sciences and engineering, and the number of researchers per million population. India's tertiary education enrollment ratio in 2019 was about 28 percent, increasing from about 15 percent 10 years earlier. This contrasts with the rapid growth in China from about 20 percent 10 years ago to over 50 percent in 2020. This ratio is less than 10 percent in Pakistan and Afghanistan, similar to the median level in Sub-Saharan Africa.[3] For other South Asian countries, the ratio is between 10 and 20 percent. As for annual doctoral degrees awarded in natural science and engineering, in 2016 India had about 10,000, compared with 30,000 in China. In 2013 (the latest year available), Bangladesh had less than 300 and Sri Lanka less than 100.[4] In contrast to China's 1,225 researchers per million population (2017), India had 253 (2018), and Sri Lanka, 106 (2015).[5] According to the Global Innovation Index, the South Asia region ranks poorly, except for India, which is ranked 52nd (China is ranked 14th).

Digitalization is profoundly changing all stages of scientific discovery and innovation. As in other aspects of the converging technology revolution, there is great potential for transformative discoveries and catch-up but also the very real risk that poor countries will fall even further behind. Historically, the transition from scientific research to adoption at scale and development impact has taken decades. Box 4.1 and figure 4.2 show the long trajectory from the discovery of improved wheat seeds, which launched India's Green Revolution, to significant reductions in rural poverty. A similar lag is evident in the discovery of knowledge about HIV/AIDS to lowering the number of deaths.

Admittedly, several sociopolitical factors affect the speed of transition from discovery to application. But digitalization is speeding up both the discovery of new solutions (for example, the speed of discovery of COVID-19 vaccines is unprecedented) and potentially their applications in society. High-speed connectivity, computing power, shared data, and advanced human capital are required, as well as a reorientation of the innovation system to address the most pressing human capital challenges—inclusiveness, sustainability, and resilience. The convergence of technologies constitutes an urgent mandate to higher education institutions in the region to adapt teaching content, research collaboration, and cross-disciplinary engagements to remain competitive.

Currently, most innovations involve new products, processes, or business models, partially enabled by digital technologies or embodied in data and software. These innovations are being driven by the large reduction in the costs of collecting and processing data and of producing and distributing knowledge and information. Data are a key input in innovation. They help businesses to explore new areas of product and service development and to adjust products to market demand. However, despite the promise, technological innovations driven purely by market considerations of profitability are not addressing the immense human capital challenges in South Asia.

As the discussions with country experts revealed (chapter 2), these technologies offer the alternative possibility of democratized, frugal innovations from the bottom up that address local needs and build community resilience. Digital innovations such as virtual simulations, 3-D printing, sensors, and data collection instruments provide new opportunities for experimentation and can be achieved at the local level using universities, community labs, and maker spaces. Furthermore, innovation is becoming more collaborative because a greater variety of expertise in different fields is required, and, with digitalization, lower-cost communication and collaboration are possible, even across countries (OECD 2019a). Innovations that help improve human capital outcomes can increasingly be generated through multistakeholder collaborative ecosystems that operate at the local level and are supported by national and even global flows of knowledge.

A related development is the increasing use of big data and artificial intelligence (AI) to increase the productivity of science. AI is being applied to all phases of the scientific process, including automated extraction from scientific literature and large-scale data collection to optimize experimental designs and even to create new knowledge. AI is also being combined with robotic systems to automate some areas of scientific research requiring intensive experimentation such as molecular biology (OECD 2018).[6]

BOX 4.1 Timeline from Scientific Discoveries to Adoption of Technologies at Scale: The Green Revolution and Treatment of HIV/AIDS

Developing and using technologies to achieve development outcomes face common challenges, including identifying and scaling workable solutions, often through global, national, and subnational exchanges of political, knowledge, and financial resources and through building conducive capacities at multiple levels of societies. It has taken decades for technologies to show impact on social indicators. Figure 4.2 traces the stages of technological adoption for two of the most significant global efforts since the 1950s.

The Green Revolution is credited with saving millions of lives from famines in India and elsewhere (figure 4.2, panel a). It dates back to the early 1960s, when American agronomist Norman Borlaug visited India, bringing with him 100 kilograms of wheat seeds he had crossbred in Mexico. Until then, Mahatma Gandhi's vision of "Village Republics" had guided the Indian government's policy. Its agricultural community development program had reached 16,300 villages by the late 1950s. Starting from about the mid-1960s, however, diffusion of the technologies characterizing the Green Revolution became the policy priority, specifically the adoption of high-yielding varieties and modern agricultural techniques such as irrigation, tractors, and fertilizers, in specific regions of the country, supported by agricultural extension and a network of agricultural colleges. By 1970, wheat yields had begun to rise. Nevertheless, the impact on rural poverty was negligible. By the 1980s, policy makers with a renewed focus on the multidimensionality of poverty had launched the Integrated Rural Development Program. Over time, environmental problems associated with the Green Revolution technologies (such as overuse of fertilizers and pesticides and depletion of groundwater)—problems not originally understood or addressed—began to offset the gains of the Green Revolution through negative impacts on health and sustainability.

HIV/AIDS is another case in point (figure 4.2, panel b). In the late 1970s, doctors in Africa observed opportunistic infections. The first HIV clinical case was identified in the United States in 1981. Antiretroviral therapy was shown to be effective in 1996, 15 years after the epidemic started. However, communities and national authorities in developing countries continued to be in denial, faced with the stigma attached to the disease. The creation of the Joint United Nations Programme on HIV/AIDS (UNAIDS) to promote international dialogue, resource mobilization, and publication and dissemination of research findings was a critical turning point in the fight against HIV/AIDS. As the Global Fund to Fight AIDS, Tuberculosis and Malaria and bilateral aid began to fill financing gaps, multilateral agencies shifted focus to strengthening capacities across relevant ministries to plan and implement national programs. It was only in 2005—10 years after the discovery of an effective treatment—that the rising trajectory of HIV-related deaths was finally reversed.

Several lessons can be drawn from these examples. The adoption of technologies occurs within a sociotechnical system, both within nations and globally. The pace of scientific discoveries, technology diffusion, and application at scale depends on sequencing and coordination of efforts to achieve political alignment and on the relevant capacities at the subnational, national, and international levels. How many lives could have been saved if these efforts had been accelerated and programs had been able to reach the poor a decade or more earlier?

Source: World Bank study team based on Global Fund–World Bank HIV/AIDS Programs (2006); India, Government of, Ministry of Statistics and Program Implementation (2015); Immerwahr (2015); Liu, Kanehira, and Alcorta (2015); Mansuri and Rao (2013); UNAIDS (2010).

FIGURE 4.2 **Timeline from Discovery to Impact at Scale: Green Revolution (1960–2010) and HIV/AIDS (1990–2010)**

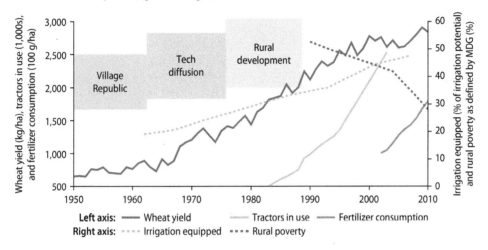

a. India's policies, agrotechnologies, and outcomes before and after the Green Revolution

Left axis: ——— Wheat yield   ┈┈┈ Tractors in use   ——— Fertilizer consumption
Right axis: ┄┄┄ Irrigation equipped   ▪▪▪▪ Rural poverty

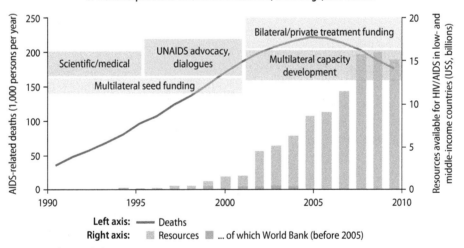

b. World responses to HIV/AIDS: Political will, knowledge, and finance

Left axis: ——— Deaths
Right axis: ▪ Resources ▪ ... of which World Bank (before 2005)

Source: Liu, Kanehira, and Alcorta 2015.
Note: Panel a: "Village Republic" refers to the vision of Mahatma Gandhi that guided the Indian government's approach to rural development, with an agricultural community development program that reached 16,300 villages by the late 1950s. "Tech diffusion" refers to the period when adoption of the technologies of the Green Revolution (seeds of high-yielding crop varieties and modern agricultural techniques such as irrigation, tractors, and fertilizers) were promoted in specific regions of India, supported by agricultural extension and a network of agricultural colleges, as well as subsidies and price policies. "Rural development" refers to the launch of the Integrated Rural Development program by the government of India, signaling a shift from the priority given to technology diffusion. Panel b: Scientific and medical research, which began with the first detection of cases in Africa and the United States, resulted in the discovery of antiretroviral therapy by 1996. The newly established UNAIDS began promoting international dialogue and resource mobilization, including from the World Bank, in 1996. Bilateral/private treatment refers to the launch of bilateral official development assistance and the Global Fund to Fight AIDS, Tuberculosis and Malaria. HIV/AIDS = human immunodeficiency virus/acquired immunodeficiency syndrome; kg/ha = kilograms per hectare; g/ha = grams per hectare; MDG = Millennium Development Goal; UNAIDS = Joint United Nations Programme on HIV/AIDS.

In the face of these rapid advances in the digitalization of science and innovation, South Asian countries must build up their capabilities in specific areas to benefit from the converging technology revolution. These include the following:

- Developing advanced digital skills, including in the fields of data analytics and artificial intelligence to make effective use of these technologies for social and economic development

- Building large-capacity digital networks to harness the large data flows and to enable more collaborative research among groups in the country and between these groups and researchers in other countries

- Facilitating access of local researchers to the expensive high-performance computers required to exploit large datasets

- Building capacity among policy makers and regulators to make informed policy decisions, including strategic investment decisions, regulations, and safeguards for the use of these technologies—among them, countries' positions on cross-border data flows, privacy, security, and agency concerns

- Adjusting science and technology policies and systems to the new context of the global innovation system driven by digital technologies

Higher education plays a significant role in addressing the innovation lag and realizing the potential of converging technologies. Convergence integrates knowledge, tools, and ways of thinking from the life and health sciences, social sciences, humanities, and arts, as well as the physical, mathematical, and computational sciences, engineering disciplines, and beyond, to form a comprehensive framework for tackling scientific and societal challenges found at the interfaces of multiple fields. By merging these diverse areas of expertise with a network of partnerships, convergence stimulates transitions from basic science discovery to practical, innovative applications.[7] There is global awareness that the convergence of technologies will require new workforce training and educational initiatives, such as the creation and expansion of cross-domain programs in universities; updated convergence competencies as part of general education requirements; cooperative research and transdisciplinary teaming to expose students to collaborative research; funding of alternative educational credentials and workforce certificates; and effective systems for lifelong learning and workplace training.

The challenges faced in developing this capacity in South Asia, with the possible exception of India (although India is significantly lagging behind the more advanced countries) are enormous. One solution is to adopt the Open Science approach, which takes advantage of the lower costs of sharing scientific and technological information, as well as the growing concerns about the governance of technology. The Open Science paradigm aims to make scientific data, information, and publications openly available to speed the attainment of social goals and allow more citizen participation in the process of developing technologies.[8] This broader engagement with society is facilitated by ICT-enabled collaborative platforms that can enhance the entire research process from agenda setting to coproduction of research and dissemination of scientific information. With the appropriate governance and institutional frameworks, Open Science can also help regional collaboration and develop local capacity in less advanced countries.

Building up strong national research and education networks (NRENs) is necessary for both Open Science approaches as well as for harnessing the power of the digital revolution in science and innovation. The South Asia region has considerable strengths in this area, as mentioned earlier. These already existing networks can offer platforms for expanded regional collaboration in knowledge development and capacity building for local innovations that can support human capital development.

Technology for Local Resilience and Community Innovation

In addition to existing human capital challenges and the devastation wreaked by COVID-19, the South Asia region is vulnerable to new shocks—among them, climate change. About 800 million people are living in areas projected to become climate hotspots by 2050, most of which are located in disadvantaged areas. By that time, an estimated 40 million Asians will be climate migrants in the face of water shortages, declining crop productivity, and rising sea levels. Furthermore, large population movements may become seasonal with changes in extreme weather, requiring frequent local adaptation.

Although private sector–led technology development has catalyzed stupendous innovations, more concerted efforts must be made to harness the new converging technologies to benefit local communities, address the needs of the poor and the marginalized, and build local resilience to future shocks. These efforts will require, in turn, a concerted effort to support dynamic, community-led ecosystems to harness converging technologies for local human development needs and to build resilience to shocks. Democratized community labs can provide public access to tools (3-D printers, laser cutters, DNA sequencers, and so on), technological know-how, and mentorship in diverse areas of expertise in engineering, programming, robotics, digital fabrication, and biotechnologies.

Examples of community-based ecosystems in South Asia include an initiative in Kathmandu, Nepal, in which local engineers were trained in AI by a Nepalese tech start-up called Fusemachines to build drones and deploy low-cost delivery systems for health equipment. Another example is the fabrication labs in Kerala, India, where students can develop problem-solving skills. These labs collaborate with the Massachusetts Institute of Technology's Center for Bits and Atoms and the global network of International Electrical and Electronics Engineers (IEEE), which have offered networking support and a dissemination platform to students to build protective medical equipment in their local start-up ecosystem. Another example is the community Mechamind in Dhaka that supports the vision of an Open Science society in Bangladesh. The team of mentors teaches underprivileged kids to work on local innovation problems and develop skill sets for future tech-based industries. The community lab has four collaborative streams, including democratized biology; AI and science, technology, engineering, and mathematics (STEM); enabling technologies; and prosthetics (Pauwels 2020).

The strength of democratized local innovation ecosystems such as community labs serves not only the inclusion agenda for human capital, but also its empowerment. Innovation in such places is driven by a common ethos characterized by interdisciplinary, peer-to-peer knowledge sharing, higher self-esteem, acceptance, and empathy. They provide an innovative and alternative incubator for individual and collective empowerment, which can supplement efforts to build trust, legal and regulatory protections, and formal governance mechanisms, which are covered in the next chapter.

Conclusions

The long-term impact of the converging technology revolution on employment is difficult to gauge and varies by sector and across countries. For the South Asia region to benefit from its demographic dividend, it will need to promote the use of digital technologies among the vast labor pool in the informal economy. Under all scenarios, an important challenge will be the need to reskill and upskill large segments of the the labor force, especially women and vulnerable populations. The demand-side shock created by the pandemic has led to massive job losses and highlighted the need for adaptive social protection mechanisms and reinsertion into the labor market. Digital platforms for jobs and online training offerings as well as technology-enabled payment of targeted social assistance need to be specifically designed for poor and informal sector workers with limited connectivity and digital skills.

The rapid advances in digitalization of science and innovation raise even bigger challenges for countries in South Asia. New research and support are needed to understand how to inspire communities to create social capital and sustain connectedness in ways that strengthen their capacity to adapt and utilize technologies for shoring up resilience. Bolstering community-created digital and public spaces is promising. Action is required on two fronts: building up the advanced skills and capabilities for innovation through cross-border collaboration and Open Science and supporting dynamic, community-led innovation ecosystems to harness converging technologies for local human development needs and to build resilience to shocks.

Notes

1. Indonesia's manufacturing productivity is twice as high as India's. The Republic of Korea's electronics and chemical industry is 18 times more productive than India's and 30 times more productive than Indonesia's.

2. The labor cost "advantages" are due to not only low wages, but also poor working conditions and lack of enforcement of labor regulations.

3. World Bank, World Development Indicators (database), http://wdi.worldbank.org/table/2.8.

4. Natural science and engineering include the physical and biological sciences, mathematics and statistics, computer sciences, and engineering (National Science Board 2018; UNESCO 2015).

5. World Bank, https://data.worldbank.org/indicator/SP.POP.SCIE.RD.P6, 2020. Data are for latest year available on the website.

6. This advance has already progressed to the stage that it is raising questions about whether machines should be included in academic citations and how to consider them in reporting patenting. See OECD (2019b).

7. Consider the ongoing commercial impact of convergence in the field of nanotechnology. With some US$1.5 billion spent annually on federal R&D in the United States, the *annual* growth in nano-enabled products has been estimated at about US$200 billion in recent years. Similar multiplier effects can be expected from convergence-enabled biomedical technologies, which are set to disrupt the national health care industry's annual expenditures of more than US$3 trillion with new products stemming from the convergence in science, engineering, and technology.

8. For a rationale for Open Science, see UNESCO, "Open Science," https://en.unesco.org/science-sustainable-future/open-science.

References

ADB (Asian Development Bank). 2018. *How Technology Affects Jobs.* Mandalyuong, Philippines: ADB.

Global Fund–World Bank HIV/AIDS Programs. 2006. *Comparative Advantage Study.* Washington, DC: World Bank.

ILO (International Labour Organisation). 2018. *Emerging Technologies and the Future of Work in India.* Geneva: ILO. https://www.ilo.org/newdelhi/whatwedo/publications/WCMS_631296/lang--en/index.htm.

Immerwahr, Daniel. 2015. Thinking Small: The United States and the Lure of Community Development. Cambridge, MA: Harvard University Press.

India, Government of, Ministry of Statistics and Program Implementation. 2015. India Millennium Development Goals. India Country Report 2015. http://mospi.nic.in/sites/default/files/publication_reports/mdg_2july15_1.pdf.

Liu, W., N. Kanehira, and L. Alcorta. 2015. "An Overview of the UN Technology Initiatives." United Nations Inter-agency Working Group on a Technology Facilitation Mechanism. https://sustainabledevelopment.un.org/content/documents/7810Mapping%20UN%20Technology%20Facilitation%20Initiatives%20July%2023%202015%20clean%203.pdf.

Mansuri, Ghazala, and Vijayendra Rao. 2013. *Localizing Development: Does Participation Work?* Policy Research Report. Washington, DC: World Bank.

National Science Board. 2018. *Science and Engineering Indicators 2018.* https://www.nsf.gov /statistics/2018/nsb20181/report. National Science Board, Alexandria, VA.

OECD (Organisation for Economic Co-operation and Development). 2018. "Artificial Intelligence and Machine Learning in Science." In *OECD Science, Technology and Innovation Outlook 2018: Adapting to Technological and Societal Disruption.* Paris: OECD Publishing. https://doi .org/10.1787/sti_in_outlook-2018-10-en.

OECD (Organisation for Economic Co-operation and Development). 2019a. *Digital Innovation: Seizing Policy Opportunities.* Paris: OECD Publishing. http://www.oecd.org/publications /digital-innovation-a298dc87-en.htm.

OECD (Organisation for Economic Co-operation and Development). 2019b. "Fostering Science and Innovation in the Digital Age." OECD, Paris. https://www.oecd.org/going-digital/fostering -science-and-innovation.pdf.

OECD (Organisation for Economic Co-operation and Development). 2020. "Technology and the Future of Work in Emerging Economies: What Is Different." OECD Social, Employment and Migration Working Paper No. 236, OECD, Paris.

Pauwels, E. 2020. "Converging Technologies for Human Capital." Paper prepared for South Asia Technology for Human Capital Study, World Bank, Washington, DC.

R&D World. 2020. "Global R&D Funding Forecast." (February): 46. http://www.rdworldonline .com.

UNAIDS (Joint United Nations Programme on HIV/AIDS). 2010. *Global Report: UNAIDS Report on the Global AIDS Epidemic 2010.* Geneva: UNAIDS.

UNCTAD (United Nations Conference on Trade and Development). 2018. *Technology and Innovation Report: Harnessing Frontier Technologies for Sustainable Development.* Geneva: UNCTAD. https://unctad.org/system/files/official-document/tir2018_en.pdf.

UNCTAD (United Nations Conference on Trade and Development). 2021. *Technology and Innovation Report: Catching Technological Waves Innovation with Equity.* Geneva: UNCTAD.

UNESCO (United Nations Educational, Scientific and Cultural Organization). 2015. *UNESCO Science Report: Towards 2030.* Paris: UNESCO. https://unesdoc.unesco.org/ark:/48223 /pf0000235406.

World Bank. 2016. *World Development Report 2016: Digital Dividends.* Washington, DC: World Bank.

World Bank. 2018. *Which Countries Are Better Prepared to Compete Globally in the Disruptive Technology Age?* Washington, DC: World Bank.

CHAPTER 5

Human Capital Empowerment: The Importance of Trust, Data Safeguards, and Protection of Vulnerable Groups

Introduction

The converging technology revolution in the human development and other sectors is enabling the collection of data on the biological, cognitive, socioeconomic, and other characteristics of individuals and population groups on a mass scale. Within the human development sector, the converging technology revolution is the most advanced at the global level in the health sector because of the integration of multiple technologies in the prevention, diagnosis, and treatment of disease. Furthermore, these technologies are used not only by health care service providers but also increasingly by private individuals, who are able to collect, analyze, store, and transfer their personal health data. Advances being made as well in the application of converging technologies in the education sector are enabling data to be collected on children and young people for everything from test performance to personality traits. Moreover, applications in other sectors that contribute to human capital, such as agriculture and water and sanitation, collect data on other aspects of human activity.

The possibility of combining heterogeneous data from different sectors of human activity and the recognition of new patterns in the data by artificial intelligence (AI) and machine learning techniques are fueling further innovation and discoveries that could accelerate human capital outcomes. Services can be customized for individuals (such as through personalized medicine or learning), for groups of people (such as through tracking of epidemics), and through data-driven decision-making and improved monitoring and accountability. However, possible innovations come with equally important risks of misuse and harm through the development of dual-use technologies, as well as greater disempowerment if citizens are not able to understand the implications of granting consent for their data to be used, or if they do not have the possibility of later revoking consent.

This chapter discusses three issues related to the governance of converging technologies that are central to the inclusion and empowerment agenda for human capital: (1) trust in the use of technology; (2) data governance; and (3) governance of converging and dual-use technologies.[1] Trust underlies the effective deployment of all technologies, but the role of trust has become amplified in the context of converging technologies. Power asymmetries, the opacity of technology value chains,[2] and the spread of digital misinformation[3] and targeting of vulnerable groups are dominating the global discourse. The role of data governance is especially important because the possibility of using and combining data from different sources lies at the heart of converging technologies. Specifically, the linking of sensitive health, education, and social protection data with digital identification can be used to advance surveillance and profiling of vulnerable populations and individuals. Protections for the collection, storage, and use of data, especially of children, women, and marginalized communities, are an important component of agency and empowerment. The specific issues related to the governance of converging and dual-use technologies, in particular AI applications, are covered in the final part of the chapter.

The Role of Trust in the Use of Technology

The *World Development Report 2021: Data for Better Lives* recognizes that a digital economy is built on trust (World Bank 2021). Trust in and between various data participants and operations increases the frequency and diversity of data transactions. Such trust is sustained by a robust legal and regulatory framework that is divided into two categories: (1) *enablers*, which facilitate access to and the reuse of data, and (2) *safeguards*, which prevent data misuse and misappropriation.

Building civic trust requires a multipronged effort to improve digital access, media literacy, regulation of the private and public sectors, institutions for enforcement, involvement of active civil society organizations in technology issues, responsiveness to citizen concerns by both the private and public sectors, and accountability for misuse, discrimination, and violence.

In the context of South Asia's deep structural inequalities in human capital and power asymmetries, trust is a major factor that complicates the deployment of technologies, in particular dual-use technologies, which, while designed for beneficial purposes, can be misused against particular population groups. The erosion of societal trust in technologies tends to affect the poor, vulnerable, and marginalized households and communities disproportionately and via a variety of channels, including data invisibility, bias, exclusion, exploitation, and cyber vulnerability.

Technologies are embedded in sociopolitical systems. Who deploys them, for what purposes, and whether for the benefit or detriment of which population groups determine whether technologies are adopted and have broad positive outcomes for social cohesion and well-being. The COVID-19 pandemic demonstrated that although digital technologies, such as contact tracing apps and health status apps, support public health strategies, the lack of concrete safeguards for data collection and the murky communication and accountability about their intended use raised suspicions and undercut their effectiveness in many countries.

A recent study that probed these issues found that trust is central to the use of technology, and it extends beyond concerns about data or privacy (Ada Lovelace Institute 2020). People need to trust that technology can solve a problem, that it will not fail, and that it can deliver outcomes reliably. Moreover, any application linked to personal identification is viewed as a high-stakes one and may invite misgivings that technological systems that rely on information such as biometric markers could end up discriminating against individuals or groups based on certain categories. Similarly, business models based on "people are the product" philosophies are increasingly contested by those who resent being categorized and then productized. Digital tools must proactively protect against harm. Legitimate fears of discrimination or prejudice must be addressed from the outset by applying "trust-by-default" and "trust-by-design" principles (Consumers International 2019), and thus in-depth public deliberation must be built in from the start. Apps are judged to be part of the overall system (and not just the specific technology) in which they are embedded such as institutional policies and procedures, data architecture and analytics, outreach and explainability, and accountability and redress mechanisms. Finally, tools are viewed by the public as political and social phenomena as opposed to neutral "objects." Technology cannot be isolated from questions about the nature of the society it will shape—solidaristic or individualistic, divisive or inclusive. In short, a normative framework enshrining the values that the technology and its deployment embody is as important as the tools themselves.

DIGITAL PLATFORMS AND MISINFORMATION

The COVID-19 pandemic revealed the danger of misinformation, undermining public trust. Misinformation spread first and foremost through digital platforms. Although this phenomenon is evident worldwide, it seems especially pronounced in South Asia. For example, real-time global analysis of disinformation about the pandemic showed that India is one of the largest sources of false information (Poynter 2021). A recent study by the London School of Economics of WhatsApp use in India, the country's predominant messenger application, found that civic trust is a derivative of ideological, family, and communal ties (Banaji and Bhat 2019). The study also found that prejudice, in addition to digital illiteracy, is a primary concern about disinformation, and thus is central to the use of social media.

INVASIVE PRIVATE AND PUBLIC DATA COLLECTION SYSTEMS

Increasingly, invasive data collection systems are gathering sensitive information about identities, beliefs, behaviors, and other personal details, using capabilities for probing, monitoring, and tracking people and even influencing their behavior. Even potentially riskier is the role of data brokers who aggregate datasets from a variety of different sources (including scraping public records and buying or licensing private data) to brand or create profiles of individuals or specific population segments that are then sold for a profit.[4] Left unregulated and operating outside of the public's purview, such practices can incite mistrust. Such risks extend to threats to privacy. These trends can have especially serious implications for women and children and the participation of traditionally disadvantaged and marginalized groups, including specific castes, tribal populations, and nonbinary gendered communities, as well as linguistic, religious, and ethnic minorities. The European Union's General Data Protection Regulation (GDPR) specifically prohibits the processing of "personal data revealing racial or ethnic origin, political opinions, religious or philosophical beliefs, or trade union membership, as well as genetic data, biometric data for the purpose of uniquely identifying a natural person, data concerning health or data concerning a natural person's sex life or sexual orientation" (European Union Agency for Fundamental Rights and Council of Europe 2018). The characterization of these data as "sensitive" is based on discrimination law and the principle of upholding human rights. However, researchers have documented that data-driven initiatives in the Global South cannot escape "concerns of discrimination, bias, oppression, exclusion and bad design" that are "exacerbated in the context of vulnerable populations, especially those without access to human rights law or institutional remedies" (Arun 2019).

Safeguards against such misuse are particularly critical in the health and education sectors. Digital health platforms and diagnostic tools used by both medical institutions and individuals also scoop up enormous amounts of information. The rise in wearable devices is one example. Although the prospect of being able to automatically track sleep patterns, calories burned, blood pressure, heart rate, and even how an individual's body reacts to a drug through ingestible sensors (Thompson 2013) has potentially great benefits, individuals, especially in developing countries, currently have little knowledge about where their health and medical data are stored, who has access to them, and whether they can be sold and used by other parties.[5]

Similar concerns beset online educational systems. The COVID-19-induced explosion in distance learning has fueled concerns that school systems using digital platforms may not fully understand the data privacy provisions when signing up for these services, even though the platforms use data analytics to monitor performance in real time and students contribute videos and other highly sensitive data (Bloomberg Law 2020). A recent report suggests that data breaches in educational institutions in the United

States tripled in 2019 and resulted in "the theft of millions of dollars, stolen identities, denial of access to school technology and IT systems for weeks" (CISO MAG 2020).

EXCLUSION OF "DATA INVISIBLE" GROUPS

The flip side of invasive data collection systems is that those without digital access (that is, those with no connectivity or devices) or who lack digital skills will not figure in the characterization of the population and their needs. As indicated earlier, there are deep inequalities in digital access and use between and within households in South Asia. An overreliance on "automatic" data collection methods can exclude these highly vulnerable groups and further undermine trust. Such exclusions may exacerbate biases that limit the effectiveness and validity of AI algorithms trained on easily accessible data, reinforcing the need for greater transparency in the use of data.

"Data invisibility" is a corollary of the digital divide across South Asia and is likely to affect traditionally marginalized communities such as tribal communities, castes, religious and linguistic minorities, and migrant workers. In an increasingly digital world, data invisibility also means limited voice and reinforces restrictions on effective participation in social, economic, and political spheres. For example, according to one estimate, 60 percent of the 10 million most popular websites on the internet are in English.[6] Hindi at 0.1 percent is the top South Asian language, with other major regional languages such as Bengali and Urdu not even appearing on the list. Thus even linguistic majorities in the region are at a great disadvantage, while relegating the linguistic minorities to near oblivion online with significant consequences.

ABUSE AND EXPLOITATION OF WOMEN

A recent study covering Bangladesh, India, and Pakistan found that over 70 percent of the sampled women "regularly contended with online abuse, experiencing three major abuse types: cyberstalking, impersonation, and personal content leakages," which led to "emotional harm, reputation damage, and physical and sexual violence" (Sambasivan et al. 2019). Even though public awareness of cyberviolence against women is rising, and it is increasingly a punishable offense, the offense is rarely fully redressed by imposing fines or sentencing the perpetrator (Vidhisastras 2020). Because of social stigma and widespread fear of reprisals, less than 2 percent of women reported the abuse to either the police or the platform (Sambasivan et al. 2019).

Harassment of women online reinforces the violence and abuse faced by women in South Asia. The region already has the highest rates of gender-based violence in the world, which is a barrier to economic participation (World Bank 2020).[7] Online harassment also prevents women from participating in the digital economy. Deliberate harassment excludes women, especially from marginalized communities, from civic and political participation.[8]

SPECIAL VULNERABILITY OF CHILDREN

The unauthorized collection of data from children is pervasive even in advanced countries. With the great advances being made by private education technology companies in the South Asia market, this issue should be a high priority for regulation. The widespread collection of cognitive, emotional, and behavioral data through adaptive learning tools and affective computing algorithms, as documented in the case studies conducted for this study, presents dangers to children—see chapter 3 based on the background paper by Pauwels (2020). The potential for online child sexual abuse is extremely high in South Asia.[9] Meanwhile, online harassment and cyberbullying in school require additional regulations and enforcement mechanisms, but regulations may be difficult to enforce because many of the offenses may occur outside of school. Imposing an appropriate punishment that does not permanently damage the defendants, who are often minors themselves, is a challenge worldwide. A Bangladesh teenager recently developed a promising alternative for which he was awarded the International Children's Peace Prize. The teen's app, Cyber Teens, allows young people to report cyberbullying confidentially through a network of volunteers and social workers. Victims can send a screenshot or call recordings to the organization and receive legal or technical support (*Tennessee Tribune* 2020).

Some recent examples highlight how unregulated private actors can undermine trust in the use of technology through unauthorized collection and use of children's data. A recent complaint in the United States claimed that YouTube had violated the Children's Online Privacy Protection Rule[10] by collecting personal information—in the form of persistent identifiers used to track users across the internet—from the viewers of child-directed channels without first notifying parents and getting their consent. YouTube monetized these data by delivering targeted advertisements to viewers of these channels (FTC 2019). TikTok has similarly been the target of complaints by advocacy groups that it has not honored previous commitments to stop collecting names, email addresses, videos, and other personal information from users under the age of 13 without a parent's consent (Singer 2020).[11] These global platforms are widely used in South Asia, where legal safeguards and parental understanding and oversight are much weaker. In an encouraging development, high courts in India and Pakistan's Draft Personal Data Protection Bill (Lexology 2020; *National Law Review* 2020) recognized the "right to be forgotten" (RTBF), a concept that has been upheld in Organisation for Economic Co-operation and Development (OECD) countries. RTBF is the right to have personal information removed from publicly available sources, including search engines, databases, and websites, under certain limited circumstances. Enforcing this right for children who may experience negative repercussions from online content they posted is especially important because of their vulnerability and the resulting potential long-term mental and emotional consequences.

Data Governance

Erosion of trust is a global leitmotif of the digital age. As just illustrated, digital contracts and interactions among citizens, communities, and consumers, on the one hand, and public and private organizations, on the other, are increasingly shaped by the rights and safeguards pertaining to the collection, use, sharing, and governance of data. Other aspects of data in the human development sectors also deserve attention.

DATA RIGHTS

The "rights-based" approach to data enshrines the principle that personal data are not owned. Instead, they are an extension of oneself. The objective of this approach is to reverse the power asymmetry between data subjects and the third parties profiting from their data. An individual may give consent for the use of his or her data, or use of the data may meet the criteria for a third party to "control" personal data for a particular use. From the perspective of human capital empowerment, the question of what happens to the data being collected, shared, and analyzed is critical. Data collection often occurs involuntarily, imperceptibly, or in the course of routine digital activities, but with pro forma "consent" as a prerequisite for accessing services and goods.

Thus terms such as *data colonialism* have begun to surface in the policy literature as international firms collect and "own" or "appropriate" vast amounts of data about a country's citizens and other assets (Coleman 2019; Couldry and Mejias 2019). This practice runs the risk of the country and its citizens becoming data-dependent on foreign entities able to extract value and exercise influence and power through the data.

This issue is particularly relevant to tribal and indigenous populations, which too often have been subjected to data collection as an instrument of oppression. Data initiatives must take special note of the historical circumstances and legal rights of these groups. Issues to note include ownership and use of data related to the communities themselves and to their cultural and environmental assets. Indigenous data sovereignty (IDS)—"the right of Indigenous peoples to control data from and about their communities and lands, articulating both individual and collective rights to data access and to privacy"[12]—is a model framework for maximizing the benefit of open data for Indigenous peoples and other users of Indigenous data.

The issue of data rights will also affect the cultural assets of linguistic and other minorities. Although they face daunting disadvantages in the online world because of their limited presence, there is also the danger that without adequate protections their knowledge can be appropriated and used more widely.[13]

A highly charged issue and an important area for policy development is the use of children's data. Children's earliest digital identity is typically created by others, well before the children themselves understand its implications or can provide their consent. In most cases, children are unaware of the implications of their digital participation because of their own digital illiteracy and the inscrutability of digital platforms. In response, some data protection and consumer protection regimes have created a heightened obligation for digital platforms to ensure the lawfulness of children's consent. However, the approach must be balanced in practice with freedom of online speech.

Legal protections for individuals and population groups, especially those that are vulnerable and historically disadvantaged with limited access to enforcement of their rights, should focus on the issues of proportionality, purpose limitations, and data minimization. For example, is the pervasive collection of cognitive, emotional, and behavioral data from children through so-called affective computing algorithms necessary for successful education? Are ubiquitous facial recognition, drone monitoring, and sentiment analysis required for specific purposes that can be clearly delineated? Such protections are essential for building the trust of communities that technologies are being deployed for purposes that are beneficial to them.

DATA ETHICS

Data ethics has become a major policy area as understanding of both the positive and negative potential of data-driven decision-making, especially via AI, has expanded and the need to address the moral dimensions of data has become more acute. Data ethics is critical in the health, education, and social protection sectors in which vulnerable populations have relatively less autonomy and are exposed to significant harm if data processors mishandle their information (Wade 2007). Crafting meaningful ethical data guidelines for use by practitioners requires avoiding the tendency to overlegislate to address every possible instance of misuse, which may deter innovation, while providing advice on essential safeguards. Furthermore, the incorporation of data ethics into education and training curricula should receive greater attention going forward.

THE CURRENT STATE OF DATA PROTECTION IN SOUTH ASIA

In general, South Asia's legal and regulatory environment for personal data is underdeveloped and lacks implementation and enforcement. For example, some countries, including Afghanistan, Bangladesh, India, and Pakistan, lack comprehensive data protection legislation.[14]

Even so, progress is being made. Nepal and Sri Lanka have enacted comprehensive data protection legislation and created special categories of "sensitive" data, which receive heightened protection. These categories typically include information related to ethnicity, political affiliation, biometrics, and genetics. Other countries are in the process of enacting comprehensive data protection legislation. Personal Data Protection Bill 2019 is currently pending in the Indian parliament. In Pakistan, such a bill was introduced in its parliament in 2018. These bills typically create user rights and privacy protection provisions such as a right to consent to all treatment of personal data, a data minimization policy, a limitation on the uses of collected data, and a right to require correction of information. The provisions of the bill in India also include protections for children, gender, and caste.

Although the South Asia region has through its acts, bills, and high court decisions made progress on data protection, critical trust-building provisions remain unaddressed, including on data misuse and misappropriation. Furthermore, many of the laws contain exceptions for government surveillance and so-called emergencies, enabling the executive to wiretap communications with a relatively low threshold. For example, in certain jurisdictions, or for certain crimes, only a low "burden of proof" standard is needed. Simple suspicion of a connection with legal wrongdoing is usually enough.

Government and private bodies in South Asian countries continue to lack the institutional preparedness needed to address and enforce data policies systematically, consistently, or fairly. Data protection authorities, where they exist, tend to have a limited mandate and capacity. India, for example, currently has no national data protection authority, and concerns have been raised that the body proposed in the Personal Data Protection Bill may struggle to build internal capacity, leading to either under-regulation or overregulation. Under-regulation would defeat the intent of the bill, whereas overregulation would add unnecessary burdens for compliant businesses. The bill also does not provide adequate checks and balances to ensure that the central government and the data protection authority exercise their vast supervisory powers in a reasonable manner (Burman 2020).

Collectively, the specific needs of and protections required by vulnerable populations, including women, children, marginalized communities, and minorities, must be addressed. Limited government awareness of the role of data and related policies in digital transformation, as well as the government's limited capacity to develop such policies, mean that vulnerable populations are at risk, and the data illiteracy of those groups hinders their participation in demanding and designing such policies.

Governance of Converging and Dual-Use Technologies

Converging technologies pose special governance challenges posed by (1) their dual-use properties, whereby they may have been designed for beneficial purposes but could also cause harm to populations; (2) their heavy reliance on large combined datasets; and (3) the frequent use of AI analytics in support of delegated decision-making. The extension of these technologies and the use of the data collected ostensibly for other purposes to carry out state surveillance or discrimination against minorities pose a serious risk. Responses to the COVID-19 crisis have accelerated large-scale behavioral surveillance, collection of personal and medical data that could be used subsequently for other purposes, monitoring of movements of the population, and screening of activity and contacts on mobile phones. For example, recently India powerfully upgraded its technological surveillance capacities to deploy individual facial recognition at railway stations and airports, algorithmic crowd analysis during street protests, and mobile, contactless biometrics identification for temperature detection (BiometricUpdate.com 2020). Beyond state surveillance is a whole range of risks related to manipulation of information and behavior modification for commercial surveillance purposes.

The risks posed by converging technologies as applied to human capital are presented in table 5.1, grouped by existing sources of risk and those likely to emerge in the next few years. Vulnerable populations, in particular, are exposed to risks of cybercrime,

TABLE 5.1 **Risks Posed by Converging Technologies**

Timeline	Risk stratification
Current	*Data commodification* • Commodification of behavioral, emotional, and biometric data of children and other vulnerable populations for education scoring, future commercial targeting, and exclusion/discrimination schemes *Failure of technological design and predictive value* • Biases in datasets and algorithmic design, as well as poor performance in predictive value, may lead to system (access, delivery, optimization) failures, with corrosive implications for underserved groups *Manipulation for state surveillance* • Commodification of behavioral, transactional, socioeconomic, and consumption data for social credit systems and exclusion/discrimination schemes • Use of personal data to silence civil society resistance, repress traditional media structures, and harm the reputation of knowledge institutions, leading to the closure of virtual civic spaces, affecting people's resilience and society's social fabric *Information disorders, disinformation, and hate speech* • Use of personal, demographic, ethnic, behavioral, and emotional data collected on children and adults for targeting disinformation and polarization, for emotional manipulation and hate speech, and for radicalization • Mobilization of large population subgroups around violent narratives, including around elections
Near term to five-year time frame	*Cyberoperations, cyberbullying, and social engineering* • Use of personal and emotional data for social engineering, leading to more efficient and more powerful acts of cybercrime • Use of biometric data for precision biometric attacks (cyberattacks where autonomous malware uses soft facial, voice, or biometric features for impersonation) • Exfiltration of sensitive datasets about populations to direct attacks to vulnerable subgroups (such as targeting groups facing food insecurity and retaliating against specific minorities, based on biometric data) • Automated data poisoning—that is, poisoning data in critical information infrastructure such as that related to medical or hospital databases or biometric, civic, and electoral registries • Cyberattacks targeting automated supply chains, thereby affecting food security and the delivery of essential human capital services • Cyberattacks in which autonomous malware weaponizes other dual-use technologies (such as biotech, 3-D printing, and robotics, including drone technologies)

Source: Adapted from Pauwels 2020.

cyberbullying, and social engineering, which are enabled through data collection efforts across human development sectors.

The discussion around the appropriate governance structures for converging technologies is still in the early stages as societies and governments struggle with understanding the full ramifications of the ongoing technological revolution. However, a few priority areas can be identified for urgent work to promote the inclusion and empowerment agenda for human capital in South Asia. Developing regulatory standards and legal mechanisms for ensuring transparency, accountability, and local empowerment can help to ensure that these principles are integrated into the design and deployment of converging technologies from the outset and that their beneficial potential can be harnessed for human capital development.

TRANSPARENCY: PROTECTING FAIRNESS AND SAFEGUARDING AGAINST BIAS

Algorithms used for decision-making, whether for personalized learning, job selection, or medical diagnostics, can harbor biases, conscious as well as unconscious, based on the training datasets used and the design of the algorithms themselves. Two sources of bias are of special concern. The first is datasets that may in fact accurately reflect society's existing biases, thereby pointing to unresolved societal problems that need to be addressed. For example, a society's discrimination against a particular group cannot be corrected by "fixing" the underlying dataset or algorithm. Instead, it requires a societal solution. The second is datasets that mispresent reality. In such datasets, because of a low participation rate or inaccuracies in synthetic data, the data and the algorithmic application layer fail to accurately reflect reality. It has been shown, for example, that many current facial recognition algorithms fail to discern the features of African Americans, particularly women, because the underlying data did not take into account a representative sample of faces. Another example comes from the new field of emotion analysis, in which facial recognition analysis struggles to detect the "smiles" of Asian women. Such algorithms now play a critical role in programs such as those measuring stunting. Use of such algorithms can perpetuate or worsen the existing inequalities in a society where there is insufficient data on the marginalized and vulnerable or on the biases of the developers.

To correct for missing data and economize on the substantial cost of creating and labeling new datasets, a frequently employed second-best solution is to rely on synthetic datasets.[15] This work-around can introduce new problems, however. For example, drawing on the case study on the diagnosis of stunting noted in chapter 3, images and measurements of a few real human beings are used to create images of artificial individuals through feature-based editing, yielding people with different heights and weights and features such as double chins and bony structures (Pauwels 2020). Without further scrutiny, this approach introduces uncertainty about its performance when applied to real individuals.

Transparency thus requires public scrutiny of the training dataset and whether the predictive analyses produced by algorithms are accurate, precise, and reproducible. Is the training dataset large enough, of good quality, and representative of the target population? Were data labeling and data curation handled with competence? Was the algorithmic bias systematically assessed? How often will the algorithm and dataset be updated? Has documentation of the model learning process been prepared according to acceptable standards? Proposals for the verification and validation of datasets and algorithmic decisions may find growing resonance in the sphere of public policy (such as that related to court decisions and social welfare programs) in a bid to build and restore the public's trust in government. For private organizations, this approach will likely meet resistance about disclosing confidential information and trade secrets. Although some private organizations will be motivated by self-interest and reputational concerns and may allow measures such as third party auditing, the majority will be guided by the enactment of principle-based legislation and upholding of robust ethical principles and standards.

Several countries in South Asia have developed national AI strategies and intend to use AI in the health and education sectors, but this approach should be carefully assessed (box 5.1). It is unlikely that these countries can elucidate the type of

---

BOX 5.1 National Artificial Intelligence Strategies in the South Asia Region

India (2018). India's National Strategy for Artificial Intelligence, a work in progress but not fully funded or implemented, focuses on the use of technologies to ensure social growth, inclusion, and elevation of the country to leadership in artificial intelligence (AI) on the global stage. Strategically, the government also seeks to establish India as an "AI garage," incubating AI that could be applicable to the rest of the developing world.

NITI Aayog, the government think tank that wrote the "in-progress" national AI strategy report, calls this approach #AIforAll. The strategy aims to (1) equip and empower Indians with the skills to find quality jobs, (2) invest in research and sectors that can maximize economic growth and social impact, and (3) scale Indian-made AI solutions to the rest of the developing world. Areas for AI interventions include health care, agriculture, education, smart cities and infrastructure, and smart mobility and transportation. The section "Ethics, Security, Privacy and Artificial Intelligence" highlights the need to be conscious of the factors of the AI ecosystem that may undermine ethical conduct, impinge on one's privacy, and undermine the security protocol. Budget allocation for Digital India, the government's umbrella initiative to promote AI, machine learning, 3-D printing, and other technologies, was almost doubled to Rs 30.73 billion (US$477 million) in 2018 (Bhattacharya 2018).

Pakistan (2018). Pakistan announced an AI initiative in April 2018 to be funded at US$3.3 million over three years. The project will be supervised by the Higher Education

(Box continues on next page)

BOX 5.1 **National Artificial Intelligence Strategies in the South Asia Region**
(continued)

Commission (HEC), and six public sector universities were selected to develop nine AI research labs (Khan 2018). In addition, the government announced the Presidential Initiative for Artificial Intelligence and Computing to develop human capacity.

Sri Lanka (2018). Sri Lanka, through its National Export Strategy Advisory Committee, announced the launch of an AI Nation to promote the education of 5,000 data scientists from 2018 to 2025 (Daily FT 2018). This measure will serve as a step toward drafting an AI national plan.

Source: Pauwels 2020.

---

safeguards just discussed. Such safeguards are especially urgent in relation to children and education. The United Nations Children's Fund (UNICEF), in partnership with many institutions, has developed a new memorandum on AI and children's rights that recommends developing a framework for AI based on child rights and delineates rights and corresponding duties for developers, corporations, parents, and children around the world (UNICEF 2019). Collaboration with international partners and civil society organizations is needed to develop the appropriate transparency standards for AI and converging technologies.

ACCOUNTABILITY FOR MISUSES OF TECHNOLOGY

As converging technologies become more automated and more decentralized, there is an accompanying lack of clarity about who will be held accountable for their potential and actual misuses. Furthermore, the technological supply chain is long and complex, involving training data, data centers, cloud-based computing services, fiber-optic networks, and highly specialized technical expertise, which is often distributed across the world. These weakly regulated supply chains of AI create pervasive cybersecurity and data security threats and a growing accountability gap, with vulnerable populations exposed to considerable harm and without recourse to legal remedies. Some damage could be irreversible and affect human rights.

Because reliance on corporate ethics or self-regulation will not be sufficient, rules-based accountability mechanisms and institutions are required, but that will take time. Thus there is an urgent need to develop and operationalize a "theory of no harm," based on a normative framework, and to undertake sociotechnical system analysis, which would contribute to the development of a strong governance framework that empowers human capital (Pauwels 2020). Such approaches could include, for example, the right to object to automated decision-making and a moratorium on the use of facial recognition technology and other methods for which currently legal protections are weak or cannot be enforced.

BUILDING DOMESTIC CAPACITY IN CONVERGING TECHNOLOGIES FOR
REGULATION AND CIVIL SOCIETY OVERSIGHT

Because large technology and data firms dominate the supply chain of converging technologies, an important question is how local capacity will be built. The knowledge and capacity of local firms, innovators, researchers, and regulators are of critical importance. Leaving aside the potential diffusion of knowledge through private sector firms, strategies such as Open Science and community innovation labs can also contribute to building a critical mass of dispersed knowledge about the good uses of converging technologies and their potential risks. University programs that focus on the gender, equity, and ethical implications of AI and converging technologies could help, along with building up the specialized technical knowledge required to deploy these technologies. In this way, such values could be built into technological design, deployment, and governance.

Conclusions

Technologies operate within a social system, and the sociotechnical system includes values, norms, culture, laws, and regulations. Trust underpins the equitable deployment and use of technology, but it must be reinforced with laws and regulations, as well as the ability to enforce them. Trust has to be built at the community level through informed participation in technology governance and through the ability to seek redress of grievances. Without informed consent and the opportunity for grievance redressal, the risks of biases that discriminate against vulnerable populations, misinformation, violence against women and marginalized communities, and human rights violations cannot be mitigated.

The system for regulation and governance to deal with converging technologies and the paradigm shift in the nature of data and its uses have yet to be put in place globally. In South Asia, the data and technology governance framework is still nascent or almost nonexistent in some countries, and it needs to be quickly strengthened or developed to enhance human capital empowerment and inclusion in the context of deep-rooted structural inequalities. Policy safeguards on data in the human development sectors are urgently needed to address data proportionality, data purpose, and data minimization and to include issues of where, how, and for how long data are stored and how they can be used. Special protections are required for children, minorities, women, and groups that are often excluded from the mainstream. Furthermore, in addition to laws and rules, institutional mechanisms for enforcement, accountability, and redressing grievance must be established, and the capacities of local communities and vulnerable groups (who often lack basic literacy skills) must be built. Establishing a fair and transparent data and technology governance system is challenging, but it can also be transformational, not only by empowering vulnerable groups to participate in a rapidly changing society, but also by enabling the exploitation of multiple sources of data through AI to enhance innovations in service delivery and build resilience to future shocks. Finally, normative foresight scenarios, audits, and impact assessments can help to prevent, mitigate, or establish accountability for causing harm to, or acting on biases that discriminate against, vulnerable populations.

Notes

1. The chapter does not cover all issues related to data or data governance, including the economic use of data. These issues are comprehensively analyzed in the 2021 *World Development Report* (World Bank 2021). This chapter focuses on issues especially relevant to human capital empowerment in the South Asia region.

2. For example, the Federal Trade Commission accused Facebook in December 2020 of buying and freezing out small start-ups to choke competition. Google is being sued by the US Department of Justice for using anticompetitive tactics to boost its advertising monopoly by acting as an intermediary for both ad buyers and sellers.

3. Between 2018 and 2020, Facebook and Twitter announced that they had taken down 147 influence operations (Goldstein and Grossman 2021).

4. According to *World Development Report 2021*, "Some 4,000 data brokers operate worldwide in an industry valued at US$250 billion" (World Bank 2021).

5. For a discussion of collecting, storing, and sharing personal health data, see Estrada-Galiñanes and Wac (2020).

6. W³Techs, https://w3techs.com/technologies/overview/content_language.

7. As indicated in the World Bank's South Asia Human Capital Plan, surveys in selected cities show the extraordinarily high levels of abuse. In six cities in Bangladesh, 84 percent of women face verbal abuse and sexual remarks in public spaces. In Lahore and New Delhi, over 90 percent of women face harassment on public transport or feel unsafe. Over 85 percent of Afghan women report at least one form of gender-based violence.

8. A recent study in India found that women politicians received an average of 113 abusive tweets a day (substantially more abuse than women politicians in the United Kingdom and United States) and that "women politicians from marginalized communities were disproportionately targeted." Muslim women politicians received 94 percent more ethnic or religious slurs than women from other religions. Women politicians who are Scheduled Castes, Scheduled Tribes, and Other Backward Classes received 59 percent more caste-based abuse than women from "general" castes (Amnesty International 2020, 6, 21).

9. The United Nations (2009) estimates that more than 750,000 online predators are exchanging material on child sexual abuse, streaming live abuse of children, encouraging children to produce sexual material, or grooming children for sexual abuse.

10. US Federal Trade Commission, Children's Online Privacy Protection Rule (COOPA), https://www.ftc.gov/enforcement/rules/rulemaking-regulatory-reform-proceedings/childrens-online-privacy-protection-rule.

11. In a settlement, Musical.ly, now known as TikTok, agreed to pay US$5.7 million to settle a US Federal Trade Commission allegation that the company illegally collected personal information from children. They have since been accused of leaving videos posted by children online, using problematic age verification processes, and "putting children at risk of sexual predation" (Singer 2020).

12. State of Open Data, 21. Indigenous Data Sovereignty, https://www.stateofopendata.od4d .net/chapters/issues/indigenous-data.html.

13. Ideally, linguistic diversity should be a source of strength. Unfortunately, it mostly ends up being a source of discrimination and marginalization (Agnihotri 2019).

14. This review did not include Bhutan and Maldives.

15. The cost of curating a customized dataset can amount to 75 percent of the overall cost required for developing an algorithm.

References

Ada Lovelace Institute. 2020. *No Green Lines, No Red Lines: Public Perspectives on the COVID-19 Technologies.* https://www.adalovelaceinstitute.org/our-work/covid-19/covid-19 -report-no-green-lights-no-red-lines/.

Agnihotri, Rama Kant. 2019. "Linguistic Diversity and Marginality in South Asia." In *Handbook of Education Systems in South Asia,* edited by P. Sarangapani and R. Pappu. Global Education Systems. Singapore: Springer.

Amnesty International. 2020. *Troll Patrol India: Exposing Online Abuse Faced by Women Politicians in India.* https://decoders.blob.core.windows.net/troll-patrol-india-findings /Amnesty_International_India_Troll_Patrol_India_Findings_2020.pdf.

Arun, C. 2019. "AI and the Global South: Designing for Other Worlds." In *The Oxford Handbook of Ethics of AI,* edited by Markus D. Dubber, Frank Pasquale, and Sunit Das. Oxford, UK: Oxford University Press.

Banaji, S., and R. Bhat. 2019. "WhatsApp Vigilantes: An Exploration of Citizen Reception and Circulation of WhatsApp Misinformation Linked to Mob Violence in India" (blog). November 11, 2019. https://blogs.lse.ac.uk/medialse/2019/11/11/whatsapp-vigilantes-an -exploration-of-citizen-reception-and-circulation-of-whatsapp-misinformation-linked-to -mob-violence-in-india/.

Bhattacharya, A. 2018. "India Hopes to Become an AI Powerhouse, with Inspiration from China," World Economic Forum, Cologny, Switzerland.

BiometricUpdate.com 2020. "Biometrics Tenders Issued in India for Railway State Security, Govt Employee Management." October 30, 2020. https://www.biometricupdate .com/202010/biometrics-tenders-issued-in-india-for-railway-station-security-govt -employee-management.

Bloomberg Law. 2020. "'Explosion in Distance-Learning Tech Sparks Privacy Worries." April 6, 2020. https://news.bloomberglaw.com/privacy-and-data-security/explosion-in-distance -learning-tech-use-sparks-privacy-worries.

Burman, Anirudh. 2020. "Will India's Proposed Data Protection Law Protect Privacy and Promote Growth." Carnegie India, New Delhi. https://carnegieindia.org/2020/03/09/will-india-s-prop osed-data-protection-law-protect-privacy-and-promote-growth-pub-81217.

CISO MAG. 2020. "Ransomware Attacks and Data Breaches on U.S. Schools and Colleges Triple in 2019." February 28, 2020. https://cisomag.eccouncil.org/ransomware-attacks-and-data -breaches-on-u-s-schools-and-colleges-triple-in-2019/.

Coleman, D. 2019. "Digital Colonialism: The 21st Century Scramble for Africa through the Extraction and Control of User Data and the Limitations of Data Protection Laws." *Michigan Journal of Race and Law* 24 (417). https://repository.law.umich.edu/mjrl/vol24/iss2/6.

Consumers International. 2019. "Consumer IOT: Trust by Design 2019: Guidelines and Checklists." https://www.consumersinternational.org/media/239715/trust-by-design-guidelines.pdf.

Couldry, N., and U. A. Mejias. 2019. "Data Colonialism: Rethinking Big Data's Relation to the Contemporary Subject." *Television and New Media* 20 (4): 336–39. https://doi.org /10.1177/1527476418796632.

Daily FT. 2018. "Sri Lanka to Launch 'AI Nation' as Next Wave of IT Growth." August 29, 2018. http://www.ft.lk/ittelecom-tech/Sri-Lanka-to-launch--AI-Nation--as-next-wave-of-IT -growth/50-661730.

Estrada-Galiñanes, Vero, and Katarzyna Wac. 2020. "Collecting, Exploring and Sharing Personal Data: Why, How and Where." *Data Science* 3 (2): 79–106.

European Union Agency for Fundamental Rights and Council of Europe. 2018. *Handbook on European Data Protection Law.* https://fra.europa.eu/sites/default/files/fra_uploads /fra-coe-edps-2018-handbook-data-protection_en.pdf.

FTC (Federal Trade Commission). 2019. "Google and YouTube Will Pay Record $170 Million for Alleged Violations of Children's Privacy Law." News release, September 4, 2019, FTC, Washington, DC. https://www.ftc.gov/news-events/press-releases/2019/09 /google-youtube-will-pay-record-170-million-alleged-violations.

Goldstein, Josh A., and Shelby Grossman. 2021. "How Disformation Evolved in 2020." Tech Stream, Brookings Institution, Washington, DC.

Khan, D. 2018. "Govt Allocates Rs 1.1 Billion for Artificial Intelligence Projects in 6 Universities." https://propakistani.pk/2018/04/23/govt-allocates-rs-1-1-billion-for-artificial-intelligence -projects-in-6-universities/.

Lexology. 2020. "Pakistan Releases Updated Draft of Personal Data Protection Bill." May 14, 2020. https://www.lexology.com/library/detail.aspx?g=c5af9973-3e77-42e5-9f03-2402050ab764.

National Law Review. 2020. "High Court in India Reaffirms the Need for an Individual's 'Right to Be Forgotten.'" December 6, 2020. https://www.natlawreview.com/article /high-court-india-reaffirms-need-individual-s-right-to-be-forgotten.

Pauwels, E. 2020. "Converging Technologies for Human Capital." Paper prepared for South Asia Technology for Human Capital Study, World Bank, Washington, DC.

Poynter. 2021. "Fighting the Infodemic: The #Corona VirusFactsAlliance." https://www.poynter .org/coronavirusfactsalliance/.

Sambasivan, N., A. Batool, N. Ahmed, T. Matthews, K. Thomas, S. Gaytán-Lugo, D. Nemer, et al. 2019. "'They Don't Leave Us Alone Anywhere We Go': Gender and Digital Abuse in South Asia." In *Proceedings of the 2019 CHI Conference on Human Factors in Computing Systems (CHI '19).* Association for Computing Machinery, New York. https://doi.org/10.1145/3290605.3300232.

Singer, Natasha. 2020. "TikTok Broke Privacy Promises, Children's Groups Say." *New York Times,* May 14, 2020. https://www.nytimes.com/2020/05/14/technology/tiktok-kids-privacy.html.

Tennesse Tribune. 2020. "Bangladesh Teen Wins International Peace Prize for Anti-Bullying App." November 19, 2020. https://tntribune.com/bangladesh-teen-wins-international-peace-prize -for-anti-bullying-app/.

Thompson, Cadie. 2013. "Wearable Tech Is Getting a Lot More Intimate." December 26, 2013. https://www.entrepreneur.com/article/230555.

UNICEF (United Nations Children's Fund). 2019. "Artificial Intelligence and Children's Rights." https://www.unicef.org/innovation/sites/unicef.org.innovation/files/2019-06/AI%20%26%20 Children%27s%20Rights_28%20June.pdf.

United Nations. 2009. "Promotion and Protection of All Human Rights, Civil, Political, Economic, Social and Cultural Rights, Including the Right to Development: Report of the Special Rapporteur on the Sale of Children, Child Prostitution and Child Pornography." Human Rights Council, General Assembly, United Nations, July 13. https://www2.ohchr.org/english /bodies/hrcouncil/docs/12session/A.HRC.12.23.pdf.

Vidhisastras. 2020. "Cyber Stalking in India." June 16. 2020. https://vidhisastras.com/cyber -stalking-in-india/.

Wade, D. 2007. "Ethics of Collecting and Using Healthcare Data." *BMJ* 334 (7608): 1330–31. https://doi.org/10.1136/bmj.39247.679329.80.

World Bank. 2020. "South Asia Human Capital Business Plan." World Bank, Washington, DC.

World Bank. 2021. *World Development Report 2021: Data for Better Lives*. Washington, DC: World Bank.

CHAPTER 6

Technology in the World Bank's Portfolio of Human Capital Projects in South Asia

Introduction

This chapter presents an analysis of components of the World Bank's projects in South Asia related to human capital. The analysis relies on the human capital framework and technology classifications outlined in chapter 2. Because these projects were not prepared using the human capital framework or the technology classification systems developed for this report, this is an ex post analysis. Nonetheless, analyzing the portfolio through this lens enables identification of the gaps and issues that must be addressed going forward to meet the human capital objectives in the South Asia region. Box 6.1 summarizes the scope of the portfolio analysis.

The next section decomposes the technology components of these projects along the three dimensions of the study's human capital framework—that is, those that primarily serve to build and protect, deploy and utilize, or empower human capital. They were further subdivided according to the technology classification schema in figure 2.3 in chapter 2. The components related to the build and protect dimension include individual technology products and services that support service delivery and digital platforms that, in turn, support ecosystem building for delivering services. Technology components related to the deploy and utilize human capital dimension are further broken down into those that help to develop skills (such as employability skills and entrepreneurship and firm capabilities) and specialized knowledge capital. Finally, technology components related to the empower dimension of the human capital framework include policies and institutions (such as data and technology governance to minimize risks to

---

BOX 6.1 Methodology for Analysis of the World Bank's South Asia Project Portfolio for Human Capital

In the fiscal year ending June 2020, 99 ongoing projects and pipeline projects were being managed in South Asia by the World Bank's human development (HD) Global Practices (Health, Nutrition, and Population; Education; and Social Protection). The total financial commitment was US$17.6 billion. Of this universe, 69 projects with total funding of US$15 billion were selected for analysis. Excluded projects included those smaller than US$100 million and pipeline projects in the early stages of preparation—that is, those without draft Project Appraisal Documents.

In addition to the projects managed by the HD Global Practices, the review also examined 25 projects, with a total commitment of US$3 billion, from non-HD sectors—such as water and sanitation and agriculture—that involved at least one of the HD Global Practices.

These 94 projects used different lending instruments, including traditional investment projects and those with Program-for-Results (P4R) financing, as well as Development Policy Loans (DPLs). The oldest project was approved in fiscal 2013. Because they are disbursed so rapidly, only DPLs with approval dates from fiscal 2018 onward were included in the review.

All projects were disaggregated to the most granular level of subcomponents, each of which included the available lending allocation. The 94 projects had 524 project components. Of these, 205 supported technology interventions. The designation of project components as "technology components" was relatively straightforward for investment projects because the component descriptions tend to specifically mention technology. For P4R projects and "disbursement linked indicators" in investment projects, the project results framework was interpreted to attribute outcomes to the described technology inputs and activities. For DPLs, the number and relative importance of policy actions related to technology (for example, on data governance) served to inform the pro rata allocation of the overall loan to the technology component.

The total loan commitment for the 205 technology components amounted to US$6.4 billion.

Source: World Bank study team.

---

populations) and crisis preparedness and response mechanisms. A small number of components were allocated to more than one dimension because they could contribute to both. Although these categories are admittedly broad and require some judgment, they allow assessment of the composition of the portfolio within the framework adopted for this study.

The chapter then moves to an analysis from another angle, namely the level of maturity of the technology components classified as contributing to the build and protect human capital pillar. The technology maturity framework, in line with industry standards, incorporates five levels of maturity, ranging from piloting to systemic impact. The categorization of these components was based on the descriptions in the project documents.

Breakdown of Technology Components of the World Bank's Human Capital–Related Portfolio in South Asia

The portfolio review findings confirm that the World Bank has a substantial engagement in technology in South Asia, with ongoing investments of about US$6.4 billion across all three dimensions of human capital. The findings also indicate that the scope of technology interventions tends to be focused on the build and protect pillar of the human capital framework, with relatively limited engagement in the deploy and utilize and empower pillars. About US$5.3 billion is allocated to building and protecting human capital, US$1.6 billion to deploying and utilizing human capital, and US$0.3 billion to empowering human capital.[1]

Figure 6.1 shows the breakdown of the technology components under the build and protect pillar by sector and according to the technology classification schema reflected in figure 2.3 in chapter 2. The individual technology products in figure 2.3 are broken down into those for government administration, frontline service delivery agents, and beneficiaries. Digital platforms in the classification schema in figure 2.3 are divided into two subcategories: systemic ecosystem-level interventions (shown at the top of figure 6.1) and building blocks, which are ecosystem enablers (shown at the bottom of figure 6.1). Systemic ecosystem-level interventions include the implementation of data systems and the transition to government-as-a-platform, while building blocks comprise essential elements such as connectivity, unique identification, and digital payment mechanisms.

Overall, there is no obvious predominance of Global Practices (GPs). All human development GPs (Health, Nutrition and Population; Education; and Social Protection) and various contributing GPs have comparable presences, with investments supporting government administration, frontline service providers, and beneficiaries. A significant proportion of technology investments is for frontline service providers (such as schools and hospitals). Time series analysis reveals that over time the focus has shifted from general government administration to frontline service providers, especially in health and education during the COVID-19 pandemic.[2] And yet a significant share of technology investments continues to be directed at earlier-generation technology products such as management information systems. Another sizable share of technology interventions is aimed at beneficiaries, especially in health and education. However, relatively little is spent on digital platforms either at the system level or at the building block level. Together, the two levels account for US$1 billion, somewhat equally divided between them.

Compared with investments in the build and protect pillar, investments in both the deploy and utilize pillar (US$1.6 billion) and the empower pillar (US$300 million) are smaller (figure 6.2). Of the technology investments in the deploy and utilize dimension, about US$1.2 billion is directed at developing the specialized skills needed for research and development, primarily in health and education, and US$0.4 billion at developing employability skills and entrepreneurship. These allocations are partly explained by the

FIGURE 6.1 Technology in World Bank Human Capital–Related Ongoing and Pipeline Projects in South Asia: Build and Protect Pillar

US$, millions

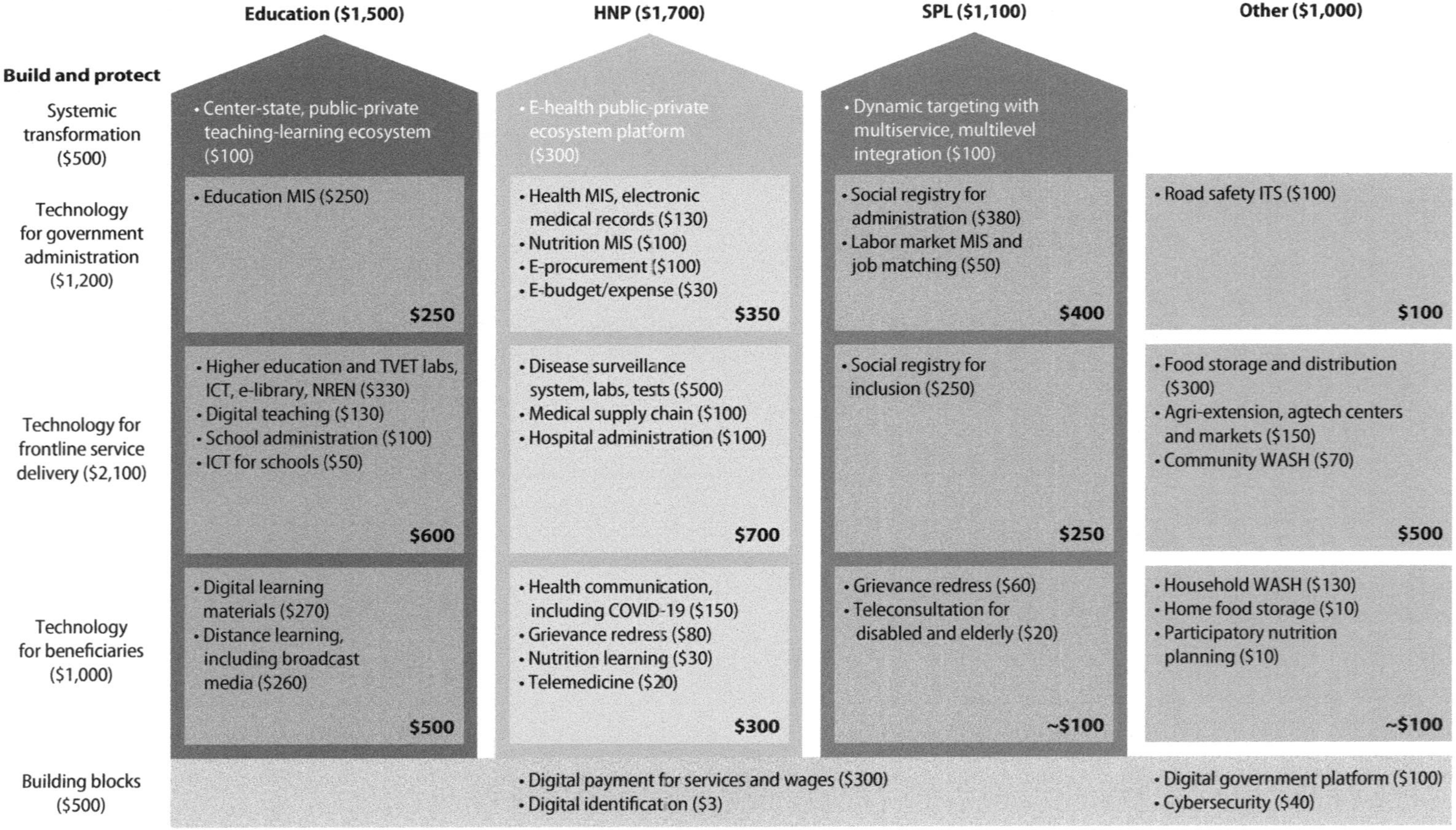

Build and protect	Education ($1,500)	HNP ($1,700)	SPL ($1,100)	Other ($1,000)
Systemic transformation ($500)	• Center-state, public-private teaching-learning ecosystem ($100)	• E-health public-private ecosystem platform ($300)	• Dynamic targeting with multiservice, multilevel integration ($100)	
Technology for government administration ($1,200)	• Education MIS ($250) $250	• Health MIS, electronic medical records ($130) • Nutrition MIS ($100) • E-procurement ($100) • E-budget/expense ($30) $350	• Social registry for administration ($380) • Labor market MIS and job matching ($50) $400	• Road safety ITS ($100) $100
Technology for frontline service delivery ($2,100)	• Higher education and TVET labs, ICT, e-library, NREN ($330) • Digital teaching ($130) • School administration ($100) • ICT for schools ($50) $600	• Disease surveillance system, labs, tests ($500) • Medical supply chain ($100) • Hospital administration ($100) $700	• Social registry for inclusion ($250) $250	• Food storage and distribution ($300) • Agri-extension, agtech centers and markets ($150) • Community WASH ($70) $500
Technology for beneficiaries ($1,000)	• Digital learning materials ($270) • Distance learning, including broadcast media ($260) $500	• Health communication, including COVID-19 ($150) • Grievance redress ($80) • Nutrition learning ($30) • Telemedicine ($20) $300	• Grievance redress ($60) • Teleconsultation for disabled and elderly ($20) ~$100	• Household WASH ($130) • Home food storage ($10) • Participatory nutrition planning ($10) ~$100
Building blocks ($500)		• Digital payment for services and wages ($300) • Digital identification ($3)		• Digital government platform ($100) • Cybersecurity ($40)

Source: World Bank study team.

Note: The numbers in each column and each row do not necessarily add to their totals because some components can be classified in two categories. Agtech = agricultural technology; HNP = health, nutrition, and population; ICT = information and communications technology; ITS = intelligent transportation system; MIS = management information system; NREN = national research and education network; SPL = social protection and labor; TVET = technical and vocational education and training; WASH = water, sanitation, and hygiene.

FIGURE 6.2 Technology in Word Bank Human Capital–Related Ongoing and Pipeline Projects in South Asia: Deploy and Utilize and Empower Pillars

US$, millions

	Education ($800)	HNP ($600)	SPL ($200)	Other ($300)
Deploy and utilize				
Skills ($400)	• Entrepreneurship ($150) • Digital skills, including TVET ($60) **$200**	• Medical technology skills ($60) **$60**		• Entrepreneurship ($50) • Firm technology capability ($70) • Digital skills ($30) **~$100**
Specialized knowledge ($1,200)	• R&D ($100) • Innovation ($190) • Higher education STEM ($220) **$500**	• Medical R&D ($160), innovation ($70) • Infection control ($300) **$500**		• Agricultural R&D ($50), innovation system ($30) • Digital innovation system ($20) • Indigenous nutrition/NRM innovation ($130) **~$200**
Empower				
Policies and institutions ($300)	• Institution for digital education policy learning ($20)	• Health emergency response capability ($20)	• Data governance ($110) • Home-based work policy ($50) • Integrated resilience ($50) **~$200**	• Digital policy institution, policy skills ($60)

Source: World Bank study team.
Note: Numbers in each column and each row do not necessarily add to their totals because some components can be classified in two categories. HNP = health, nutrition, and population; NRM = natural resource management; R&D = research and development; SPL = social protection and labor; STEM = science, technology, engineering, and mathematics; TVET = technical and vocational education and training.

relatively higher shares of tertiary and vocational training projects in the South Asia education portfolio, possibly in response to long-standing client demand. The investments related to the empower pillar, by contrast, are recent and include pipeline projects, which are in an early stage of preparation. Data governance, through prior actions related to social protection platforms in the context of Development Policy Loans, has been included in pipeline projects since the onset of COVID-19.

Given the historically strong demands for technology support by countries in South Asia with advanced capabilities, the significant human development support for more traditional information and communications technology solutions (for example, investments in back-office government administration) could be the result of legacy investments. The relatively modest presence of advanced digital solutions—for example, in public-private platforms and the use of artificial intelligence (AI)—does not align with the rapidly evolving converging technology landscape for service delivery in the region presented in chapter 3, and it suggests new areas for increased World Bank engagement. Currently, there is also limited engagement in technologically enabled solutions for the anticipated massive disruptions in employment caused by the introduction of converging technologies in agriculture, industry, and services. New approaches to building science, technology, and innovation capabilities such as Open Science and community-level innovation systems to foster resilience that leverage the power of connectivity, data, and AI are still not to be found. Furthermore, the World Bank's relatively recent limited support for the empowerment pillar of human capital is particularly striking because of the existing deep inequalities, already pervasive misuse of technologies, and the higher risks associated with converging technologies.

Assessment of Technology Maturity in World Bank Projects

The portfolio assessment also identified the extent to which the technology interventions classified under the build and protect human capital pillar allow governments to fully embark on implementing technology-enabled transformation strategies. Figure 6.3 presents a classification of technology interventions in World Bank human development projects based on a framework with five levels of technology maturity. Close to 60 percent of the assessed portfolio (by volume) is situated at stages 1 or 2, the lowest levels of maturity, which can be characterized as piloting individual technology applications or solutions. Around 30 percent of technology components are at stage 3, a level characterized as integrating and scaling multiple solutions, and only 10 percent are at stage 4, characterized as having a broader systemic impact. Examples of the latter include support for digital government platforms and financial support for the creation of public-private ecosystems. No project, however, could be classified as at stage 5, the highest maturity level, involving optimization of the use of technology at scale. Furthermore, even the relatively small portfolio classified at stages 3 and 4 is heavily skewed by sectors (highest for social protection and labor, lowest for education) and countries (India, followed by Bangladesh and Nepal).

FIGURE 6.3 World Bank Human Development Ongoing and Pipeline Projects by Level of Technology Maturity, South Asia

	Piloting		Integrating, scaling	Systemic impact	
Typical client status at each stage	Stage 1: Initial	Stage 2: Developing	Stage 3: Defined	Stage 4: Managed	Stage 5: Optimizing
Individual technology products — Government	• Predigital MIS		• Competent digital asset management		• Good governance and system stewardship
Individual technology products — Frontline	• Subsidized experiment	• Sectoral legislation • Maintenance capability	• Technology standard, skills standard	• Private participation, market formation	• Autonomous ecosystems
Individual technology products — Citizen	• Little awareness or visibility	• Municipal pilot • Sandbox approach	• Quasi-universal coverage • Pervasive user literacy	• Single-window service integration	• Real-time feedback and learning loops
Digital platforms	Initiate and expand coverage of building blocks (such as connectivity, ID, payment) →		Facilitate multiparty ecosystems and foster functional delivery platforms →		

Examples of World Bank support (health MIS):

"The project will support effective implementation of individualized monitoring and case management through a system of individual records for registered pregnant mothers, infants, and children." Bangladesh Health Sector Support Project, DLI with US$28 million

*"[Nutrition] management systems are being strengthened through innovative **mobile- and tablet-based** ICT tools that facilitate consistent, **real-time reporting and monitoring** of service delivery at Anganwadi Centers."* India National Nutrition Management Information System Additional Financing, DLI with US$52 million

*"[TB] case-based information system enables large scale monitoring, strategic purchasing, direct electronic payments to patients and providers and new adherence support technologies… will facilitate **trust, accountability, and rapid-cycle performance management.** The mix of incentives and engagement models will create an **ecosystem for private sector engagement."** India Program Towards Elimination of Tuberculosis, DLI with US$176 million

Source: World Bank study team.
Note: DLI = disbursement linked indicator; ICT = information communications technology; ID = identification; MIS = management information system; TB = tuberculosis.

Implications for Future Engagement

The portfolio review reveals that the World Bank's engagement in technology in South Asia's human capital projects is fairly substantial in total value. However, the use of technology is predominantly concentrated in the build and protect pillar of the human capital framework used in this study—that is, for improving service delivery—and positioned in the piloting stage of technology maturity. Piloting is often necessary when new technologies are being introduced, but over time scaling up is required for the technology to have an impact. A significant proportion of the technology investments in the portfolio is in traditional products such as management information systems, which may reflect the presence of older projects.

The use of technology targeted specifically at marginalized and vulnerable populations to promote inclusion is also not prominent. Many of the converging technologies mentioned in this report and already deployed by the private sector in the region are not in the World Bank's portfolio.

Finally, the work on developing public regulations and safeguards to address emerging technology impacts on data rights and market dominance in relation to human development programs is still in the early stages.

Harnessing the potential of converging technologies can help to strengthen all three dimensions of human capital in the South Asia region. Priority areas for increasing the application of converging technologies for achieving human capital objectives include optimal targeting of social transfers, at-scale delivery of customized educational content to marginalized groups, and discovery of new approaches in digital health and collaborative medical research. With pressure mounting to increase job opportunities, the rollout of digital platforms and the appropriate use of data-driven technologies has strong scale-up potential to facilitate job matching, offer reskilling and upskilling programs, and customize the delivery of social services, especially at the community level. A particularly important area for intervention is to build community innovation ecosystems that leverage converging technologies to enable the adaptation, creation, and diffusion of technologies and help build resilience to climate change and environmental degradation. Appropriate investments in promoting scientific collaboration, digital infrastructure, data sharing, and high-speed computing, possibly at a regional level, could lead to innovative ways of developing and deploying advanced scientific and technological capacity to address urgent human capital challenges. To foster inclusion and empowerment, reducing inequalities in first-mile digital access, encouraging local content, and developing policies and regulations on data and technology deployment are important areas for future World Bank engagement.

In view of the dynamic technology landscape presented in the previous chapters, the significant uncertainties posed by the COVID-19 crisis, and the wide spectrum of South Asian countries' ambitions and capabilities in the deployment of technologies, the World Bank's accelerated use of technologies will require ongoing stakeholder engagements with the private sector, local and scientific communities, and civil society. Joint learning and foresight exercises with government counterparts and external experts, focusing on the human capital challenges of each country, would facilitate the safe deployment of converging technologies for the benefit of the poor in the region.

Notes

1. The total of these amounts exceeds the US$6.4 billion mentioned earlier because about US$0.8 billion is allocated to both the build and protect pillar and the deploy and utilize pillar.

2. The time series analysis refers to a comparison of projects that were (1) approved between fiscal year 2013 and fiscal 2017; (2) approved from fiscal 2018 to fiscal 2020, pre-COVID-19; and (3) approved post-COVID-19.

CHAPTER 7

Scenario Planning: Imagining Alternative Futures for Human Development in South Asia

Introduction

Converging technologies are offering opportunities and posing threats to South Asia in human development, as well as pointing to possible pathways for addressing the socio-economic disruptions arising from the COVID-19 pandemic. Drawing on the framework outlined in chapter 2 describing the multiple interactions between human capital development and converging technologies, the study team developed a structured scenario exercise to explore a broad spectrum of uncertainties, to raise awareness of alternative futures, and to identify strategic responses. Box 7.1 describes the methodology adopted for conducting this exercise.

The framing question for this exercise was anchored in the three pillars of the human capital framework: What positive or negative roles can digital and nondigital technologies play in building and protecting, deploying and utilizing, and empowering human capital to accelerate the achievement of better human capital outcomes and strengthen societies' future resilience in the South Asia region?

The purpose of this exercise was not to arrive at a consensus about what *will* happen but to test hypotheses about what *could* happen. It was also intended to raise awareness about opportunities, hidden risks, and second-order impacts as well as to inform recommendations for improving decision-making in the use of converging technologies to develop, protect, deploy, and empower human capital.

BOX 7.1 Scenario Analysis: Uses and Methods

Scenarios are stories about what the future may be like, created through a structured process to stretch one's thinking, challenge conventional wisdom, and drive better decisions. Scenario analysis is used in situations with large multilayered uncertainties to illustrate alternative outcomes rather than one specific projection as a way of dealing with the underlying uncertainties—something that is useful when events are so unpredictable that it is impossible to attach quantifiable risk probabilities to different outcomes. The analysis helps to zero in on the drivers of change and illustrate the future differently than a projection from the past. The emphasis is on creating conceivable, relevant, and challenging scenarios to help decision-making under uncertainty. Scenarios are helpful as a set of plausible futures when used to compare, contrast, and assess whether they capture the broad range of the likeliest outcomes. By means of the resulting framework different policies can be generated and tested in alternative possible worlds.

In the first step, the study team selected four technology metatrends representing collections of digital and nondigital technologies that, with a degree of certainty, will shape human capital in the near to medium term. The metatrends are (1) the impact of converging technologies on building and protecting human capital; (2) the impact of data-driven and human-machine production technologies on the demand for and use of human capital and on economic and social structures; (3) the increasing importance of dynamic innovation ecosystems for human development; and (4) the need to develop governance arrangements for converging technologies to exploit the benefits and mitigate the risks they create for human development.

In the second step, the study team identified critical uncertainties to develop consistent alternative scenarios for the future and how they may play out. These uncertainties have the potential to tip the future in a positive or negative direction. Often two sets of uncertainties are used to develop a two-by-two matrix to trace how the uncertainties interact. This approach was adopted for this study. At the time this scenario was developed in mid-2020, uncertainties were dominated by the interplay of two factors: the trajectory of the pandemic and severity of its impact on human capital outcomes, on the one hand, and the countervailing or mitigating dynamics of cooperation versus isolation in responding to the unfolding crisis at the global, regional, national, and local levels, on the other. Many more uncertainties could be explored, thereby generating alternative scenarios. Examples are the speed required to develop and deploy converging technologies or the extent to which the world develops appropriate global governance mechanisms to reduce some of their negative impacts such as increasing inequality and loss of privacy and agency.

Participants in the scenario exercise, organized into six virtual sessions, comprised 32 experts and development practitioners representing a diverse set of experiences. They included representatives of governments; the private sector; international organizations; science and technology experts; and World Bank Group staff, representing South Asia's management team, technical staff, the human capital project, and the Technology and Innovation Lab.

This chapter goes on to describe the technology metatrends and critical uncertainties developed for this study, summarize the four scenario narratives developed with the scenario participants, and outline recommendations emanating from the scenario exercise.

Technology Metatrends

As discussed in chapter 2 of this report, the relationship between human capital and technology is close and complex and runs in both directions. On the one hand, technologies can be used to build and protect human capital and they affect the demand for human capital. On the other hand, human capital is an input in an economy's use of new technologies and contributes to the creation of new technologies. Within the broad constellation of technologies and drivers characterizing converging technologies, the study team identified four technology metatrends that were expected to affect human capital development in South Asia. Each of these metatrends is discussed briefly in what follows and is explained further in detail in appendix A.

Technologies for building and protecting human capital. This metatrend centers on the deep disruptions caused by COVID-19, which has forced millions of families to cope with extended lockdowns, school closures, health emergencies, and loss of lives and livelihoods. From a technology perspective, the pandemic has focused attention on the need for digital connectivity and readily deployable platforms, devices, and content for delivering health services, education, and social protection services. Beneath the expected flurry of new premium services offered by a growing number of private education and health platforms catering predominantly to urban households with the necessary digital connectivity and ability to pay, the day-to-day reality for most of the population remains one of digital exclusion, rising inequality, and lack of digital skills. Over time, governments will offer emergency assistance in the form of digitally enabled transfer payments and hybrid education over radio, television, and telephone, as well as telemedicine, funded by fiscal expansion, remittances by diaspora communities, and scaled-up assistance by development partners. Systemwide reforms that engage stakeholders to adapt technologies, delivery formats, and capabilities to accelerate human capital outcomes across the region are set to gain momentum over the coming years.

The converging technology revolution, however, goes beyond digital applications, as explained earlier. The potential implications for development bring into sharp relief humanity's resilience to cope with disasters, the importance of community preparedness and local support mechanisms to respond to disasters, and the rapid adoption of a range of technologies that can be used effectively in local contexts to extend and improve health and education services. The negative effects of the converging technology revolution will be reflected in a deepening of inequities for underprivileged groups without access to new technologies that rely on connectivity and digital skills, a rise in gender-based violence fueled by technology and psycho-emotional stress on families, massive learning losses for out-of-school children, and an erosion of trust in the ability of core institutions to make timely decisions based on available information.

Data-driven and human-machine production technologies revolve around the growing reliance on data-driven technologies in the manufacturing and service sectors and the impact of automation on the interactions between humans and machines. The implications range from artificial intelligence (AI)–directed automation and decision routines, to the reconfiguration of global supply chains to mitigate the risks of disruptions in the wake of disasters, to the prospect of new job requirements in the information economy. The growing role of data, both as a resource and as a source of power, is presenting challenges to traditional, export-led development pathways and is elevating the importance of connectivity and data integration among companies, employees, markets, and consumer preferences.

What are the positive and negative implications arising from this metatrend? Following patterns observed during previous technological revolutions, this period will usher in a new wave of entrepreneurial activity, accelerate the adoption of digital modes of production, and raise global awareness of the fragility of societies, with possibly long-term effects on values and consumer preferences. Potential downsides include job losses stemming from the widening gaps in technological capabilities and skills across and within societies, the growing job insecurity arising from the accelerating shift to part-time gig work, and a winner-take-all dynamic in many information markets. The latter will be driven by globally operating technology companies and data brokers that dominate data services and analytics, create markets-in-everything, and influence decision-making in all areas of daily lives. Looking ahead, the rapid emergence of converging technologies may prompt a fundamental reappraisal of traditional development models that assumed continued economic growth and a gradual shift of production patterns to emerging economies—and the mounting threats to global sustainability, which have been ignored for too long.

Dynamic innovation ecosystems for human development offer an alternative to the traditional prescriptions of centrally funded research and development programs that are out of reach for most developing countries. Instead, innovation ecosystems increasingly tap into entrepreneurial dynamism, mobilize the participation of diverse stakeholder groups, and gain digital access to expertise networks, funding, and global knowledge to create new opportunities in many spheres of human development. A tech-enabled civic culture can develop solutions to local problems through granular innovation processes and citizen-driven access to technologies, data, and bottom-up experimentation. New forms of collaboration among scientific and innovation communities are emerging that pool access to digital technologies, data, and computing power for drug discovery, personalized health and education services, and the microtargeting of social services. Countries in South Asia share a common aspiration to build up domestic capabilities to participate in this global knowledge system, assisted by global business and technology groups keen to gain a foothold in these markets and well-connected diaspora communities with access to critical know-how and funding.

What are the implications for development? The upside of these new dynamics is that governments have an opportunity to demonstrate leadership and competence in steering innovation to tackle societal challenges, mobilize resources, and build up trust, working in partnership with private companies and civil society. More granular innovation aided by digital technologies can also lead to faster learning and diffusion. At the same time, the pandemic has cast a harsh spotlight on pervasive gaps in the deployment of readily available innovations that, in turn, may raise questions about an unfinished decentralization agenda and the need to strengthen institutional capabilities for the delivery of health services, education, and social assistance, especially at the local government levels. More fundamentally, the potential for innovative technology solutions triggers questions of how digital access, usage, and skills are distributed across societies.

The governance of dual-use technologies is a question of global concern, involving issues of regulation, oversight, and data rights, with potentially far-reaching implications for human rights and dignity. The capacity to use citizens' personal data for surveillance and behavioral monitoring is expanding and will expose new cybersecurity vulnerabilities. The convergence of AI and biotechnologies, for example, has the potential to change how the genomes of humans and other species are computed, designed, and programmed. Meanwhile, the growing reliance on algorithms in many areas of life is raising concerns about the need for safeguards against biased decision-making, the concentration of data, and the spread of misinformation. At the same time, the open-source software movement is here to stay, relying on decentralized technologies and available datasets to innovate in a "permissionless" sphere beyond the purview of government control. The implications of this metatrend are subject to intense debates at the global and national levels. On the positive side, the urgency of addressing this set of issues holds out the prospect of renewed collaboration among public, private, and civil society actors to agree on a core set of normative principles to ensure equal access and transparent use, backed up by data security and inclusion for all. Greater access to AI applications and converging technologies can empower local communities to innovate with their own data and connect digitally with innovation networks elsewhere. On the negative side, the unchartered governance of dual-use technologies opens the window to encroachment, manipulation, and control, undermining trust and societal cohesion.

Critical Uncertainties

The severity of the crisis brought about by COVID-19 signals the first critical uncertainty.[3] A source of immediate concern is whether there is adequate global understanding of the transmission dynamics, the viral mutations, and the nature of the pandemic trajectories in South Asia relative to those observed in Europe and America. Would response mechanisms, including testing, therapeutics, and vaccines, be developed, accessible, and affordable in a timely manner? A related issue centers on the magnitude of the primary and secondary effects of the pandemic on lives in South Asia. Would medical facilities be able to handle the surge in demand? What are the pandemic's impacts on the existing health burden across the region? Would recurrent lockdowns and other containment measures end up reversing the region's aspirations for human development for years to come? And last, given the uncertain length and depth of the COVID-19 crisis, questions arise about the likely effects on livelihoods and equity in South Asia. Would frontline social services be able to cope and adapt? Would teachers be willing and able to teach? Would children, especially girls, return to school? Would economies reopen or revert to shutdowns? How would vulnerable groups be able to cope and survive?

This cascading set of uncertainties about the severity of the crisis engendered by the pandemic is posing fundamental questions about the trade-offs between saving lives, protecting livelihoods, and investing in recovery. On the positive end of the spectrum, the rapid production and deployment of effective vaccines would offer the prospect of a return to normalcy and underscore the critical importance of scientific capabilities and discoveries. Rapid vaccine deployment would be instrumental in containing the pandemic, showcasing well-functioning delivery capabilities across government and society at large and limiting the long-term erosion of human capital. On the other end of the spectrum, recurrent waves of infection combined with delays in preparedness and ineffective responses would exact a mounting toll across the region in terms of loss of lives, livelihoods, and prospects for recovery. Fundamentally, although this uncertainty is set against the immediate context of the pandemic, it holds broader significance in light of the region's high exposure and vulnerability to other types of disaster, especially those related to climate change.

A second critical uncertainty centers on the future of the domestic and international order, which may affect social cohesion at the national and local levels, regional political stability, and the delivery of adequate international support.[1] One source of this uncertainty lies at the national level and revolves around the impact of social distancing and economic shutdowns on vulnerable groups, the role of community leaders in helping citizens cope and maintain trust in public institutions, and the trade-offs being made in balancing the crisis response between collective safety and protection of individual rights. A second source emanates from the interactions between nations in the region and outside and involves the prospect of global tensions spilling over into the region, with impacts escalating into a rise of domestic nationalism, the imposition of trade restrictions (such as on critical medical supplies, including vaccines), or sharp downturns in foreign investments and development assistance. Reflecting the growing role of technologies in global competition and alliance dynamics, there is also lack of clarity as to whether the crisis will trigger a backlash against global technology dependence, how critical advances in science, technology, and innovation to combat the virus will be shared or used in the pursuit of diplomatic goals, and what role local innovation systems may play in developing solutions to local problems and diffusing them across wider communities.

As this set of factors grouped under the second critical uncertainty plays itself out over the course of the pandemic, a positive outcome would demonstrate the political acumen of leaders to instill trust, a deep reservoir of community resilience and compassion, and the global resolve to mobilize the needed financial support and make vaccines widely available.

On the negative end of the spectrum, the future political direction may see the region returning to isolationist policies, with some countries experiencing an increase in social polarization, and witness unfulfilled promises of regional solidarity and global support.

Four Alternative Futures

Drawing on the implications of the four technology-enabled metatrends for human development in the region and juxtaposing contrasting outcomes—positive and negative—associated with the two uncertainties, the study team generated a set of four alternative futures (see figure 7.1). A stylized summary of these scenarios follows.

Scenario A: "Strained solidarity." This is a world in which technology-enabled global solidarity is mobilized to contain recurrent pandemic outbreaks. International efforts are under way to scale up the available technologies to deliver educational services. The race for vaccines galvanizes international support. Concerted medical research collaboration is generating scientific breakthroughs. However, production bottlenecks and funding shortages delay international efforts to distribute vaccines transparently across emerging economies. Tight fiscal conditions and investor uncertainty limit opportunities for sustained economic activities. International remittances remain surprisingly strong and provide a lifeline as urban slums shut down and seasonal migrants are forced to return to rural communities. Rising demand for development assistance translates into quick-disbursing emergency programs, with social assistance programs increasingly relying on national identification programs and digital payments. Initiatives by tech companies to set up contact tracing apps are rendered ineffective and run into public mistrust about data capture and empirical validity. In the meantime, tech entrepreneurs tap into growing demand for digital education and health platforms, catering to households who can afford these services. Vacillating signals by

FIGURE 7.1 **Summary of Alternative Futures Used in Scenario Planning Exercise**

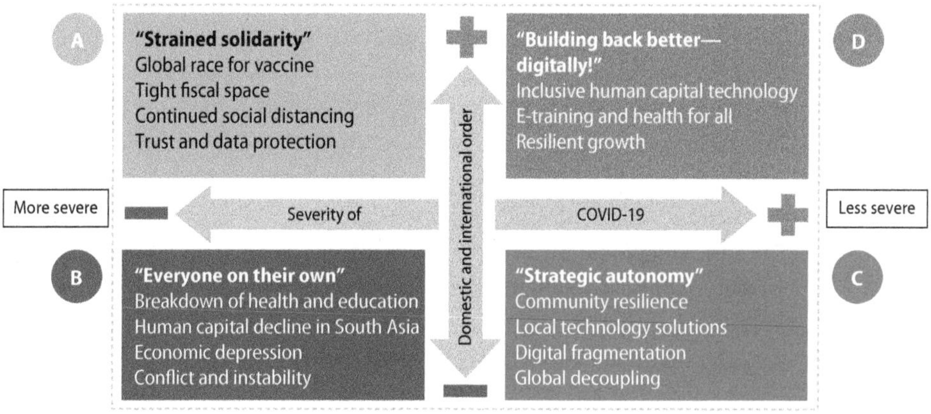

Source: World Bank study team.

political leaders undercut effective preparedness efforts, which are increasingly supplanted by self-help initiatives led by community networks and local governments. The question of who to trust is the new theme in electoral campaigns, with rising demands for digital transparency. Discussions about data protection, cross-border digital exchanges, and cyber rules are increasingly at the center of international diplomacy and national security concerns.

Scenario B: "Everyone on their own." In this scenario, initial hopes for a period of tech-enabled global solidarity to deal with the pandemic and its fallout are swept away by rising political tensions and social polarization that spill across much of the South Asia region. The recurrence of COVID-19 outbreaks and the spread of highly infectious viral mutations sweep across the subcontinent, leading to rising death tolls. Periodic flooding and other ecological catastrophes further strain an already overwhelmed health and emergency relief system. It is not surprising that the global momentum to tackle climate change loses momentum as global leaders retreat into short-term self-preservation and shut their borders to growing waves of climate migrants. Across South Asia, a precipitous drop in life expectancy and worsening malnutrition among children are evident. Subsistence farmers and informal economy workers are hardest hit, adding to a growing army of jobless adults. The educational system grinds to an almost complete standstill because of lack of funding and an inability to offer online learning for all but the most well-off, who have access to private education programs in digital exclaves. The spread of misinformation and targeted surveillance gain importance as new sources of power—only to undermine people's trust in elites and institutions, creating a volatile mix of discontent that undermines social cohesion. Poor governance results in the withdrawal of international assistance except for acute emergencies. Deglobalization and protectionism stall technology diffusion, exposing weaker countries increasingly to foreign spheres of digital dependence. The decoupling of economic activities and fragmentation of societies are compounded by a sharp drop in migration and remittances. Instability takes hold and the region descends into conflict and famine.

Scenario C: "Strategic autonomy." In this scenario, the severity of the pandemic lessens, followed by an easing of lockdown measures. However, hopes for a return to "normal" are thwarted. As Western societies question their governments' crisis responses and refuse to join ambitious new programs to prepare for future pandemics, arguments for greater regional autonomy and global decoupling gain ground. The high human toll and a slower-than-expected recovery in Western economies are countered by a resurgence of influence by East Asia. The region is reaping the benefits of early containment measures, effective vaccination campaigns, and agile production capabilities, showcasing advanced analytics and superior digital connectivity. This power shift throws the old global order into disarray and gives rise to regional techno blocs and the formation of new alliances. India's go-it-alone digital strategy serves as a model in the South Asia region, but, lacking broader engagement, it fails to overcome historical barriers to regional cooperation. National security concerns move into the foreground at the expense of ambitious social programs. Instead, local government agencies and private start-ups work with community groups to deploy their own education and health care platforms on a pilot basis. Although the reopening of health and educational services initially held out hope for postcrisis digital dividends, these local initiatives fail to scale up nationally because of lack of clarity about division of labor, standards, and funding. The poor and vulnerable continue to suffer human capital losses caused by repeated climate shocks and a low-growth trajectory. South Asia's lack of digital readiness accelerates the shift away from global supply chain networks and prevents economic catch-up. International support remains caught in a downward spiral. Fragmentation of platforms and government control over the internet prevent scale-up and international collaboration. The drop in public and private funding for science, technology, and innovation (STI) reduces the dynamism of local ecosystems. A new narrative emerges: the region's top talent is joining the global gig economy in search of a more promising future.

Scenario D: "Building back better—digitally!" In this "idealized" constellation, the pandemic is successfully contained, resulting in a modest loss of lives followed by a quick economic recovery and a technology-enabled future. In the ensuing national debate about a return to normal versus building back better, the advocates for resilience and digital transformation gain international support. Democratized, inclusive forms of innovation are thriving thanks to AI and converging technology applications, successfully meeting local needs. In the aftermath of the pandemic, new business opportunities are emerging, such as participating in supply chains and providing rural communities and migrant workers with digital services (financial, health, and educational). With access to a growing digitally literate and trained workforce, the region becomes a hub for outsourcing jobs and precision digital medicine. Schools are operating in online community spaces, attracting out-of-school students and offering personalized education and innovation training in twenty-first-century skill labs. Optimized social protection and e-health systems deliver support to vulnerable groups, expanding access for human development and economic participation. Public sector digital stacks provide essential government services with universal access and attract funding from tech entrepreneurs for additional value-added offerings. Regulatory standards for data rights and cross-border exchanges converge internationally. Thanks to broad stakeholder support for normative principles, the ability to anticipate and mitigate the risks arising from dual-use technologies is greatly improved. Propelled by the large-scale adoption of digital consumer services, South Asia's technology offerings are robust, affordable, and sustainable and attract growing interest in the Global South. As a result, the region's "resilient growth model" wins international recognition.

What are the main takeaways of the scenario exercise? Scenarios A, B, and C illustrate a variety of mounting challenges for the South Asia region unless vaccines and curative treatments for COVID-19 are available in sufficient quantities as part of ongoing international support (scenario D). Against the background of a continuing pandemic, uncertainty over the future carries grave consequences for the region in terms of lives lost and an unprecedented erosion of human capital. These outcomes could be worsened by a widening gap between the digital haves and have-nots, inadequate responses by political leaders and government agencies, and a failure to deliver promised international support.

Participants in the scenario exercise also offered a wide range of perspectives and reactions to these alternative futures. For example:

- "The recovery will be fragile and long. Health, education, and social transfers plus jobs will take center stage."

- "COVID-19 has demonstrably magnified patterns of exclusion. When inequalities become more apparent, history shows that this could trigger social movements for justice."

- "If we use tech solutions to address the needs of the poorest first, we may end up with a completely different set of services for our clients."

- "Agility and resilience will become more important than technological solutionism and specialization. Above all, being clear how we make decisions as a society over the use of technologies is crucial for determining successful outcomes."

- "Governments are beginning to realize the need for the interoperability and exchange of data. As more content goes online, this creates new responsibilities for how data are managed and who has the right to access them."

- "What is the space for global collective action and granular innovation? Can the World Bank navigate and assist in both?"

- "Can we bring these scenarios to conversations with our clients? There is a need for a national dialogue on potential futures. Decisions that may seem inconsequential today could have profound consequences in the future."

No clear consensus emerged as to the single most likely future. Participants did see evidence that several of these futures could coexist at any one time within the South Asia region, requiring vigilance of emerging trends and agile responses. This observation also supports the argument that scenarios are best used as a composite set of alternative futures (see box 7.1), thereby helping analysts to identify the scope for preemptive corrective actions by stakeholders to influence future outcomes.

Recommendations

The scenario exercise yielded the following recommendations:

- **Ensure that the World Bank tackles inequalities and digital exclusion.** In periods of heightened uncertainty and pressure to respond to crisis situations quickly and at scale, support offered by the World Bank should remain anchored in the institution's core mission: assisting vulnerable groups (such as the urban poor, women and girls, migrants, and digitally nonconnected groups) and supporting systemic reforms. To this end, World Bank support should concentrate on a handful of high-priority actions (so-called no regret measures), while drawing up actionable road maps for the medium term with top-down as well as bottom-up inputs from stakeholder groups. A key step toward restoring trust and livelihoods will be to invest in a resilient public service infrastructure and strengthen the delivery capacity at the subnational and local levels to reach intended beneficiaries, taking advantage of technologies where appropriate.

- **Leverage the World Bank's convener role.** COVID-19 has accelerated technology awareness and digital adoption to unprecedented levels globally. This is thus an opportunity for the World Bank to champion real-time learning about digital implementation experiences that have applicability beyond the pandemic (such as reaching beneficiaries at speed, devising local dashboards for crisis responses, and integrating geospatial data for mapping access routes to health services). Virtual conversations through knowledge exchanges and piloting with STI networks could be scaled up rapidly, bringing together participants from the public, private, civil society, and academic spheres. At the same time, the World Bank is well positioned to play a coordinating role among development partners to ensure an optimal degree of technology integration and reduce duplication of efforts.

- **Pursue technology partnerships between the World Bank and the private sector, scientific community, and academic and community innovators.** Partnerships with private technology companies and innovators could target specific human development programs (such as translation technologies for local content creation; remote digital diagnostic services; and health, education, and social protection programs at scale to vulnerable populations). Such arrangements would also encourage companies to contribute to development and address emerging technology governance issues. Another modality is to scout for emerging technology applications by participating in local innovation ecosystems that address the human capital needs of the poor. In addition, the Bank should pursue partnerships with key stakeholders in government, the private sector, civil society, and other international organizations to develop rules and regulations, best practices, and accountability mechanisms to improve the governance of converging technologies, thereby reducing their possible negative impacts on human capital.

- **Boost technology awareness through World Bank–organized scenario exercises.** The COVID-19 crisis has served as a reminder to governments and task teams in development agencies that speed of response matters, whether in the use of data, the rapid reconfiguration of supply chains, or the digital delivery of government assistance to intended beneficiaries. As new issues emerge, the need for ongoing guidance and feedback loops is particularly acute in technology-related areas. To respond to the growing demand for targeted knowledge support, the Bank should consider expanding just-in-time access to global expertise and participation in operationally oriented communities of practice such as the Institute of Electrical and Electronics Engineers (IEEE). Scenario exercises at the country or sector level[2] provide flexible formats in which to raise technology awareness, engage in dialogue, gather available evidence on what works and what does not through rapid policy reviews, and explore alternative responses. Human-centered design workshops are useful for co-creating new projects and interventions with clients.

- **Expand the role of World Bank's human development (HD) advisory services.** HD's advisory services, technology expertise, and partnerships are critical for laying the groundwork for a digitally enabled rebuilding program post-COVID-19. The World Bank's ongoing *Digital Economy Assessment for South Asia* (World Bank, forthcoming) and its *World Development Report 2021: Data for Better Lives* (World Bank 2021) open the door for engagements with policy makers on adapting regulatory and digital policy frameworks on data protection, dual use, and innovation, among other things. Such efforts should be accompanied by policy reviews and HD programs to support alternative technology pathways and government initiatives to accelerate human capital formation.

- **Anticipate and mitigate emerging technology risks.** The rapid evolution of technology creates new wealth and political constellations, with the potential for deepening social divisions. It also requires a new agility by government agencies and the public at large to remain vigilant against hidden risks. New approaches may involve adaptive procurement policies to avoid single-provider or solution lock-ins; new requirements for transparency and explainability of algorithms to avoid bias and data exclusion errors; systematic attention to support interoperability of data and platform stacks through open-source solutions; and actions against misinformation campaigns and cybercrime.

Notes

1. The vertical uncertainty axis in figure 7.1 is stylized and assumes convergence or lack of convergence at the local, national, and international levels, whereas each of these dimensions could have been treated independently of one another.
2. For example, although this scenario analysis was conducted on an aggregate level for the South Asia region, the World Bank's Technology and Innovation Lab used a novel technology foresight approach to raise awareness about alternative technology futures in collaboration with the Bangladesh team and clients.
3. As stated in chapter 1, this study was completed before the second wave of the pandemic in South Asia.

References

World Bank. 2021. *World Development Report 2021: Data for Better Lives.* Washington, DC: World Bank.

World Bank. Forthcoming. *Digital Economy Assessment for South Asia.* Washington, DC: World Bank.

CHAPTER 8

Accelerating Human Capital Outcomes in South Asia: The Technology Agenda

Synthesis

Devastating losses in human capital wrought by the ongoing COVID-19 pandemic in South Asia, through both its direct effects on health and its second-order economic impacts, need to be reversed urgently. Already performing poorly on many critical human capital outcome indicators and riven by deep structural inequalities, the region has to shift gears if it is to put itself on a new trajectory of equitable development.

Converging technologies, propelled by innovations in many fields and increasingly leveraging the power of artificial intelligence (AI), could transform the human capital scene in South Asia. Imagine how mass illiteracy could be eradicated if pedagogically appropriate, work-relevant resources and microlearning content in local languages were loaded on mobile devices. Literacy volunteers, agricultural extension workers, or "edu-preneurs" could then teach tens of millions of working adults using AI-supported tools. Or imagine how sensors embedded in household toilets could generate diagnostic test reports to reveal common health conditions, while information collected from sewage in community sanitation facilities could help improve public health disease surveillance in urban areas.

These solutions are no longer in the realm of science fiction; they are eminently feasible. They can help to build and protect human capital rapidly and use it productively in the economy. But consider whether data from the results of tests might be used to score individuals for health insurance or jobs. Or whether disease prevalence patterns in some communities might lead to people's exclusion from civic life. Or whether AI-powered literacy programs might be offered only to those deemed "capable" of learning. The threats to inclusion and empowerment are just as real as the potential for building human capital.

The adoption of converging technologies in human development (HD) sectors is already under way in South Asian countries. Much of the innovation is being led by the private sector for use by private health services and education providers, but there are examples from the public sector as well. In Bangladesh, community health workers have used AI to know where and when to be present for birth deliveries, thereby helping to boost neonatal survival rates. In India, efforts to accelerate the diagnosis of stunting rely on a combination of mobile cameras to capture body scans, cloud-based data processing, augmented reality, and AI for facial and body recognition. "Personalized" learning apps, which draw on AI-based tools, are being used by millions of children to identify and address individual learning needs. Public digital platforms in health and education are helping to connect frontline providers and beneficiaries with technical resources, while digital payment mechanisms are being leveraged for social assistance programs. And local-level innovations are springing up to build community empowerment and access to new technologies.

Despite these advances, the first wave of the COVID-19 pandemic in South Asia revealed major shortcomings in the use of digital technologies in the crisis response of the public sector to deliver services. These shortcomings resulted from the lack of affordable connectivity, devices, relevant content in local languages, and data privacy and safeguards, especially for vulnerable groups. At the same time, limited leadership capacity to deploy technologies and lack of digital and technical skills by both frontline providers and beneficiaries represented critical barriers to adoption. A stark example is in education, where most children in public schools have had limited or no access to education for months, resulting in high learning losses and possibly a high number of dropouts, especially among girls. Meanwhile, those in private schools and universities have ensured learning continuity by shifting to online and mobile learning.

The converging technology revolution is engulfing the world of work and innovation. Although there is uncertainty about which sectors and industries in South Asia will be affected, as well as the scale of the impact, data-driven technologies and automation will increasingly create disruptions in employment and change the demand for skills. Scientific discovery and technological innovation, fueled by data and AI, will require advanced capabilities in skills and infrastructure, while also creating the possibilities of mass participation in innovation at the local level through global connected networks for knowledge sharing.

But converging technologies also pose considerable risks for vulnerable populations in South Asia. These technologies can displace human capital in production and disempower people. They can deepen existing inequalities—gender, community, religious, and socioeconomic—through lack of access to the technology (including digital access), lack of local language content, and lack of advanced digital skills, as well as through deliberate reinforcement of biases and targeted exclusion. They also can accentuate existing power asymmetries and undermine trust and social cohesion through massive, automated disinformation, often leading to violence. In short, those who are invisible and voiceless can become more so in a digital world. Gender disparities in even basic digital access remain very pronounced. Harassment, violence, and disinformation against minorities and women are rampant in the region.

One of the most critical challenges for protecting and empowering human capital in South Asia is the lack of strong policy and regulatory frameworks, legal protections, accountability mechanisms, and supporting institutions for vulnerable populations to protect them from misuse of their data and from state surveillance, bias, exclusion, harassment, and violence. Safeguards on the use, storage, and reuse of data in the health, education, and social protection sectors are needed, in addition to general regulations on data. AI algorithms for predicting behavior and "classifying" people must be open to public scrutiny by knowledgeable peers in the same way that, for example, biases and stereotypes in textbooks are currently assessed. And yet the capacity of governments to anticipate and assess the full societal and distributional impacts of these technologies is generally limited by both technical capacity and political will.

The World Bank's portfolio of ongoing and pipeline human capital projects in the South Asia region (as of June 2020) incorporates a significant allocation for technology components (US$6.4 billion of US$15 billion in project commitments). However, using the human capital framework developed for this study, an overwhelming share of these technology components falls in the category of building and protecting human capital through improved service delivery. A relatively small proportion falls in the category of deploying and utilizing human capital to prepare for technological changes in the workplace and for adapting and developing new technologies. And even less comes within the empowerment dimension of human capital to safeguard against technology risks and proactively protect vulnerable sections of the population. About 60 percent of technology investments are at the pilot stage, with less than one-third directed at scaling technology adoption and only 10 percent at systemic transformation. No technology intervention could be classified as reaching the most mature stage of technology optimization. Equally important, technology interventions in the World Bank project portfolio do not include many converging technologies that are already part of the technology landscape for service delivery in South Asia. Areas currently not part of the technology interventions in the human capital projects of the World Bank include inclusive digital access so marginalized populations can use technology-enabled services; development of local content; design of public digital platforms to serve these beneficiaries; efforts to build the capacity for data-driven decision mechanisms using geospatial and other sources; inclusive, transparent, and accountable use of AI; and a data and technology governance framework to protect the population, especially the most vulnerable groups, from the risks of dual-use technologies.

The converging technology revolution is transformational and affects virtually all aspects of economic and social interactions. Although the broad trends of this convergence revolution are clear, multiple futures are possible for human capital outcomes, depending on how economic, social, and political factors interact with technology trends.

The scenario exercise, undertaken with a broad array of internal and external experts and relying on four technology metatrends and two critical uncertainties developed by the study team, yielded important insights about four "alternative futures" for human capital in the region. The most optimistic scenario—relatively low health, social, and economic impacts stemming from the pandemic; global support for the region; and domestic resilience—envisages a future with inclusive technology for human capital, improved service delivery for the poor, new jobs and local innovation through resilient economic growth, and empowered human capital. The other three scenarios portray a range of outcomes with varying degrees of challenge and, at the outer spectrum, outright pessimism. The latter reflects the results of the devastating impacts of the pandemic and rising geopolitical tensions with a breakdown in international solidarity and internal social cohesion, giving way to a bleak future with a breakdown in service delivery, reductions in human capital outcomes among the poor, and economic depression. These scenarios are still playing out in the region.

The key recommendations of the scenario exercise were to step up efforts to tackle inequalities and digital exclusion; leverage the World Bank's convener role around human capital–centered technology applications; pursue technology partnerships with the private sector, the scientific community, and academic and community innovators; boost technology awareness through scenario exercises with external and internal experts; expand the role of the World Bank's HD advisory services through policy reviews of technology for human capital and engagement in the regulatory framework; and anticipate and mitigate emerging technology risks, especially in relation to the use of data and AI, to reduce the exclusion of and harm to vulnerable populations.

Discussions with regional experts, outlined in chapter 2, highlighted the major cultural shift in governments and among the population in their willingness to adopt digital technologies in the delivery of human development services. Where there was community preparedness and concern for disadvantaged social groups, the willingness and ingenuity to deploy and adapt technologies to develop local solutions for resilience were on full display. Strengthening the foundations of such resilience and future adaptability is urgent because of the enormous risks to human capital in South Asia of technology-induced disruptions in employment, climate change, and environmental degradation. This urgency goes beyond stand-alone "technology interventions" and calls for building a long-term leadership capacity to deploy technologies for equitable human capital development and local solution capabilities. Above all, it calls for efforts to strengthen the social infrastructure based on trust in the use of technologies.

Based on these findings, this study avoids recommending specific technologies because their maturity, sectoral specificity, and readiness for adoption differ across countries and sectors. Moreover, there are large uncertainties in how the technology metatrends will play out in the region. Instead, this study identifies key areas in which governments, development partners, the private sector, and local communities can act to accelerate building human capital. These action areas, described in the next section, aim to improve both public and private service delivery to address human capital challenges and inequalities, strengthen inclusion and empowerment, and build resilience and adaptability, thereby enabling further innovations.

Nine Action Areas for Leveraging the Converging Technology Revolution to Improve Human Capital Outcomes

The framework used in this study elucidates a complex relationship between technology and human capital. Technology can *accelerate the building and protection of human capital* through different stages of the life cycle by means of health, education, and social protection services, as well as other contributing sectors. *The deployment and utilization of human capital* in the economic system is affected by technology, but it also shapes the future use of technology. Technologies deployed in the production of goods and services disrupt employment and alter the demand for education and skills. At the same time, specialized human capital is required to help adapt and create the new technologies important for human development and other productive sectors. Equally important, whether technology contributes to greater equity in outcomes and the *empowerment of human capital* is critical because of the tendency of new technologies to deepen inequalities in their initial phases and because of the dual-use nature of many converging technologies and thus the potential for discrimination and exclusion.

Climate change and environmental degradation in South Asia pose new threats for human capital outcomes in South Asia. Mass displacement of populations, loss of incomes, the rise of vector-borne and other infectious diseases, and the perils to human health stemming from environmental pollution constitute a new spectrum of shocks that affect vulnerable populations.

The study team has identified nine action areas that would help to leverage the converging technology revolution to accelerate human capital outcomes along its three dimensions, with a strong focus on inclusion and empowerment (figure 8.1). These nine actions can be broadly categorized as improving service delivery, building future resilience and adaptability, and promoting inclusion.[1] This categorization is intended to highlight the primary function of the action area, as there are clearly overlaps between the categories.

These action areas can also be classified according to their potential impacts, ranging from essential and cross-cutting to transformational. Cross-cutting actions enable impacts across all three functions. A second group of actions enable customization and integration for impact at scale. A third set of actions is more difficult to achieve but can have profound transformational impacts on all aspects of service delivery, building resilience and promoting inclusion.

The suggested action areas are not meant to be addressed in a mechanistic or stand-alone manner. Many are interdependent, and some may need to be undertaken first for others to be more effective. In addition, many of the proposed follow-up actions include building skills and capabilities among policy makers, frontline workers, targeted beneficiaries of government services, and the population at large, as well as specialized institutions to deal with some of the issues raised.

These action areas, as shown in figure 8.1, are intended to signal a departure from past approaches that tended to focus on piloting or scaling up specific technologies. This study argues that a broader approach will exploit the full potential of converging technologies while minimizing risks. Fully aligned with the study's human capital framework, these action areas are intentionally horizontal in nature and offer new opportunities for stakeholder involvement and codesign by combining top-down engagement with bottom-up contributions. Some actions may be easier to achieve based on the current trajectories of digital development, and some may be aspirational but are aligned with global trends for inclusive, human-centered deployment of technology. An overriding concern running across all action areas is how to build trust at the community level through transparent and inclusive decision processes, the cogent use of data and

FIGURE 8.1 Nine Action Areas in Which Technology Can Build and Protect, Deploy and Utilize, and Empower Human Capital

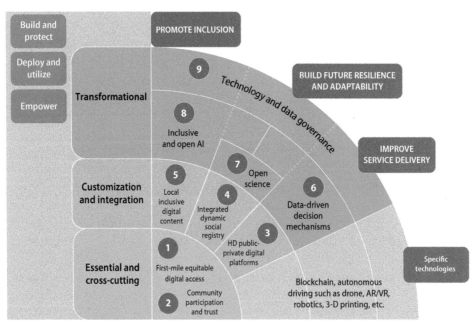

Source: World Bank study team.
Note: AI = artificial intelligence; AR/VR = augmented reality/virtual reality; HD = human development.

scientific evidence, and yet-to-be agreed-on safeguards against technological over-reach. Ultimately, the success of using this menu lies in adapting it to country-specific circumstances, strengthening capabilities among government agencies, and forging connections among the different layers and component pieces.

ACTION AREA 1: FIRST-MILE EQUITABLE DIGITAL ACCESS

Rationale, scope, and relevance

To benefit from digital technologies, all communities and individuals, especially the most vulnerable and those living in remote areas, need affordable broadband connectivity. Although usually treated as the "last mile" by telecom operators, reaching these groups should in fact become a first-mile priority. Where it is not possible to connect all households or individuals, community-based access will ensure continuity of basic services and disaster responses. Apart from connectivity, equitable digital access includes availability of affordable devices and foundational digital skills—and intrahousehold inequalities in digital access for women need to be addressed. It also includes public policy and government leadership to ensure interoperability and to ensure that key horizontal issues are addressed upfront, with the explicit understanding that the goal is to enable convergence for human development.

Proposed follow-up actions

- Construct digital road maps to ensure first-mile digital access for all.

- Pursue digital access and connectivity for all as an "all-of-government" priority. This priority would include enabling the use of AI, big data, and other disruptive technologies.

- Give the relevant government agencies the task of defining meaningful access standards (covering connectivity, price, devices, and skills) and monitor inequality in access at the local and household levels.

- Build awareness for and support the acquisition of foundational digital skills across the educational system and in daily life.

- Establish dedicated funding and transparent implementation arrangements for a universal access fund to ensure that underserved groups are being reached and connected.

ACTION AREA 2: COMMUNITY PARTICIPATION AND TRUST

Rationale, scope, and relevance

Local ecosystems bring together stakeholder groups—such as the science, technology, and innovation community, youth, local entrepreneurs, government, and venture funds—to innovate, pilot, coordinate, share, and scale up for a common purpose that is broadly understood. The ready availability of technology, open-source platforms, and data-enabled communities allows broad participation in innovation processes. Bottom-up innovation can produce locally adapted solutions. Experience has shown that local innovation capacities are critical for building resilience at the community level to prepare for and respond to crises such as natural disasters and pandemics. In times of crisis, community participation can mobilize faster, act with greater precision, offer socially appropriate responses and feedback, and operate in an environment of accountability and trust. Community involvement in the deployment of technology solutions often produces new local jobs and essential improvements in quality of life.

Proposed follow-up actions

- Apply lessons from successful community participation movements (such as that in Kerala) to crises elsewhere in the continent.

- Engage the global disaster response and management community to expand assistance and collaborate with South Asian communities on adaptation and mitigation strategies.

- Pursue opportunities to engage directly with local ecosystem clusters and community labs. This engagement can take a variety of forms, including codesign, facilitating technology transfer, financing accelerators and start-ups (possibly through challenge funds), strengthening entrepreneurial skills, or providing visibility for innovations.

- Expand local to global network links through professional communities and participation in shared solution platforms such as the Institute of Electrical and Electronics Engineers (IEEE) and Engineers Without Borders.

- Align community-enabled technology approaches with national standards and create dissemination-adaptation-feedback mechanisms to avoid unnecessary duplication of efforts and encourage scaling up.

- Build trust and social cohesion through governance norms and systems based on inclusion, transparency, and civil society participation. Over the mid- to long term, codesign of technology solutions through community-level participation can help reduce underlying social tensions and inequalities.

- Integrate technology training and reskilling initiatives with local ecosystems and forge linkups with local industries and educational institutions as part of the broader rollout across the community.

ACTION AREA 3: PUBLIC-PRIVATE DIGITAL PLATFORMS FOR HUMAN DEVELOPMENT SERVICES

Rationale, scope, and relevance

Public digital platforms (which can be built in partnership with the private sector) enable the government to provide new or better services and solutions. By combining the catalytic role of private innovations with inclusive solutions, public platforms can serve as an "equalizer" of the inequality-expanding effects of many technology applications. The design and management of platforms require a user-centric, holistic approach that focuses on the needs, contexts, and constraints of the intended beneficiaries (for example, connectivity, devices, power, and skills). This user-centric approach also encourages the participation of key stakeholder groups in the ecosystem (including the private sector) and promotes adherence to data interoperability and international standards for all platform layers. A small share of the World Bank's technology portfolio in human development in the South Asia region is currently devoted to supporting platforms with a systemic impact, with variations by country and sector.

Proposed follow-up actions

- Develop the public sector's capacity to operate platforms on its own or in partnership with the private sector, including the capacity to manage contracts and ensure safeguards on market power, data use, and stakeholder involvement.

- Drawing on the findings of the technology landscape review, address two priority areas: (1) strengthening the public sector's capacity to operate digital education platforms and (2) developing a digital health strategy and implementation road map.

- Explore private sector partnerships for digital platforms that allow rollout, customization, and scalability, recognizing that with high levels of job losses and persistent economic disruptions across the region in the wake of COVID-19, such platforms can deliver targeted learning content, skill retraining programs, and cross-training for employees and job seekers.

- Mobilize external advisory services and financial support to assist governments in building the policies and skills necessary for developing and managing digital platforms.

ACTION AREA 4: INTEGRATED, DYNAMIC SOCIAL REGISTRY

Rationale, scope, and relevance

Social protection systems directly support poor and vulnerable people to smooth consumption; reach intended beneficiaries with good targeting techniques, governance, and verification mechanisms; and generate positive outcomes for the accumulation of human capital and household resilience. Typically, programs draw on social registries to provide a variety of social services, disaster assistance, housing, cash transfers, and so on. These registries can be scaled up rapidly in times of crisis by increasing coverage or benefits, replacing cash with food where markets fail, setting aside conditionalities, and ensuring vulnerable groups have access to benefits by expanding points of last-mile delivery. Dynamic social registries also involve a rethinking of center-state/province and local government relations in the context of schemes prepared and financed at the national level with local-level implementation.

Proposed follow-up actions

- Expand portable service and income support for the large urban informal sector and returning migrants.

- Coordinate and integrate the multitude of preexisting social registries at the ministry, department, and program levels as part of the social protection architecture.

- Integrate national identification systems with social protection systems and programs and accelerate the rollout of digital payment channels, while preventing profiling and surveillance. Depending on countries' readiness, different combinations are possible, leveraging the capacities of the private sector with public sector accountability.

- Design and build inclusive, citizen-friendly, technology-enabled platforms for dynamic targeting.

- Expand the use of digital payments for government-to-person social assistance programs.

ACTION AREA 5: LOCAL INCLUSIVE DIGITAL CONTENT

Rationale, scope, and relevance

Digital content is any content that exists in the form of digital data, can be stored, and includes information that can be digitally broadcast, streamed, or contained in computer files. Creation of digital content should serve a specific purpose and be adopted, utilized, and trackable. Language is just as important in building human connections online as it is offline. South Asia is home to several hundred local languages, and yet Asian language content is almost completely absent online. This inequality of available online information in different languages has implications for who and what get represented—and by whom. Access to the internet offers an opportunity for linguistic empowerment, to translate (and adapt) important information, to share teaching resources for marginalized groups, and to create virtual communities for speakers. Translation technologies offer one solution for bridging the online language divide, while also opening new markets and jobs.

Proposed follow-up actions

- Build up local data repositories.

- Create local language access protocols to build trust and local involvement before granting permissions to use personal data.

- Enrich digital platforms with locally inclusive digital content (such as health and educational services) and opportunities (such as jobs).

- Encourage the curation and use of digital content in local languages and build the necessary skills for content creation.

- Engage with digital entrepreneurs and tech firms on how they can meaningfully contribute to local content development and use.

- Use local language content as means of community consultations and empowerment, especially with vulnerable groups.

ACTION AREA 6: DATA-DRIVEN DECISION-MAKING

Rationale, scope, and relevance

Two technologies—geospatial technologies and the Internet of Things—have wide-ranging impacts and can support human development in achieving better targeting and decision-making. Geospatial technologies enable combining spatial data with statistical and demographic data for real-time analysis of conditions on the ground and modeling of complex and dynamic scenarios. Devices that produce geospatial information include mobile phones, credit cards, and global positioning systems, which can be combined with remote sensing data for planning, monitoring, and responding. Internet of Things devices rely on sensors to collect and network data from equipment, infrastructure installations, and other "things" to help with monitoring, predictive maintenance, supply chain logistics, inventory control, and service delivery in support of optimizing real-time decision-making.

Proposed follow-up actions

- Introduce the targeted deployment of data-driven decision support mechanisms to identify and target services for population groups and in geographic areas facing human capital challenges (such as malnutrition and illiteracy) and to improve the efficiency of service delivery.

- As part of a broader digital transformation program, integrate proposed changes with existing workflows, building strong support among frontline workers and ensuring backward integration with management processes. Support digital leadership and culture change in a sustained manner.

- Promote horizontal and cross-departmental technology approaches to guide community-level interventions and facilitate data sharing across government agencies in support of targeted human capital outcomes such as nutrition.

- Leverage ongoing digital development strategies at the country level to strengthen digital policy skills and institutions and foster government data stewardship, such as between medical records and geocoded health facilities and of medical supply chain data.

- Support the transition to dynamic social registry platforms, which are relying on high-resolution satellite imagery for granular poverty estimation as in, for example, Sri Lanka.

- Undertake long-term capacity-building programs across key human development ministries—health, education, and social protection and labor—focusing on the data analytical skills of leaders and staff and improving data-driven decision-making processes.

ACTION AREA 7: OPEN SCIENCE

Rationale, scope, and relevance

Open Science is an approach aimed at enabling all countries to benefit from the digital revolution in science and the ongoing converging technology revolution. Developing countries with weak scientific capacity are at risk of being left behind both in contributing to and in benefiting from scientific knowledge. Open Science complements the open data movement and enables countries to move to the frontier of scientific and technological innovation, whether by building local innovation capacity, resilience, and adaptation or by co-creating technology and data governance frameworks as scientists, researchers, and policy makers learn about these technologies. The data revolution and new tools such as machine learning and AI are revolutionizing the process of scientific discovery— that is, "data-driven" science has joined "hypothesis-driven science." These new tools can address challenges that are inherently complex, including climate change, disasters, and sustainable development. Data, computing power, connectivity, and advanced data science skills are required to benefit from this revolution. Meanwhile, Open Science can create efficiencies of scale in the planning, procurement, and provision of data, computing power, and advanced data science skills through collaboration and shared capacities. This approach will especially benefit the smaller countries in South Asia.

Proposed follow-up actions

- Develop protocols and appropriate governance structures for Open Science to build and share the relevant scientific capability (such as advanced skills, data repositories, computing resources, or AI tools), collaborate on research using shared data, ensure the integrity of results, and communicate them in a meaningful way that builds trust in science.

- Strengthen national research and education networks and encourage country research collaboration and training on priority thematic areas.

- Incentivize collaboration through funded collaborative programs focusing on pressing challenges in South Asia such as stunting, health, food security, education, productive employment, social protection, climate change, and resilience.

ACTION AREA 8: INCLUSIVE AND OPEN ARTIFICIAL INTELLIGENCE

Rationale, scope, and relevance

AI is a general-purpose technology that is likely to transform economies and societies. It has great potential for personalizing learning and health, thereby contributing to immense improvements in quality of life. AI could optimize systems, automate processes, and improve the efficiency of service delivery. The growing use of AI in manufacturing and services could affect the level and composition of employment and alter the development pathways of emerging economies. Concerns about substitution of low-skilled labor and deepening inequalities need to be balanced with the prospects for creating new (mostly) service-related jobs, especially in urban centers. On the downside, AI can worsen inequality, bias, exclusion, concentration of power, and surveillance. However, most ethical issues remain unaddressed. AI policy making by governments plays a critical role in setting directions and accelerating development and adoption.

Proposed follow-up actions

- Engage with global AI networks and related communities, as well as local civil society groups and the private sector.

- Convene and align key stakeholders among government agencies, the private sector, and civil society groups to codesign a national road map and accelerate AI development and adoption.

- Encourage the development of inputs to inform AI strategies in the human development sectors, focusing on including measures to promote transparency and accountability and oppose discrimination and bias.

- Build up appropriate governance capabilities by leveraging and adapting policies and ethical principles developed globally to local contexts.

- Widen access to data while regulating its use to safeguard consumers, workers, users, and citizens against risks.

ACTION AREA 9: TECHNOLOGY AND DATA GOVERNANCE

Rationale, scope, and relevance

Data underpin the ongoing technological revolution. Core HD activities related to health, education, social assistance, and gender tend to lag those of other sectors (such as finance and e-commerce) in digital data protection, capture, and use. The potential benefits of using data to drive sectoral strategies and programs are falling short of reaching the majority of the population, especially in the developing world, and yet people are exposed to the risks and dangers stemming from the unethical and criminal use of data. Data sourced from the human capital sectors are especially at high risk of unethical use. The South Asia region is one of the largest data markets globally, and so the leading technology companies are drawn toward developing artificial intelligence and related technology applications.

Proposed follow-up actions

- Develop the policy and regulatory framework for data across HD sectors with the goal of promoting their beneficial use, while creating adequate safeguards for data collection, protection, storage, and use. Balance risk safeguards with incentives for innovation.

- Determine which datasets should be made available as a public good and what specialized data institutions and oversight mechanisms should be put in place or strengthened.

- Systematically apply analytical tools to data value chains, with special attention to converging technologies and the political economy over data competition and control.

- Develop standards for accountability, transparency, and grievance redress for risks to human capital agency and empowerment.

- Encourage open-source solutions to support diffusion, adoption, and transparency.

- Undertake systematic monitoring and evaluation and peer review.

Rising to the Challenge

The convergence revolution is transformational and is affecting all elements of economies and societies. As the scenario exercise made clear, much greater awareness is needed on the part of the World Bank, development community, developing country policy makers, private sector, and citizens at large about both the opportunities and challenges posed by the convergence revolution.

The deployment of converging technologies for human capital is bound to soon pick up speed. Because of the rapidly changing technology landscape, the World Bank should seize the opportunity for participatory technology foresight and undertake scenario planning exercises with government agencies, the private sector, and community groups engaged in innovation to hear different voices and to sensitize all participants to what is at stake. The World Bank should step up its policy dialogue and operational support for data management, governance, laws, regulations, implementation mechanisms, and ethics specifically related to health, education, and social protection. This policy orientation, embedded in a theory of no harm, should be accompanied by operational mechanisms for transparency and accountability in using converging technologies. The World Bank is well positioned to encourage and facilitate regional collaboration among stakeholders in order to develop a consensus on and share approaches to innovations in the use of technology for human capital.

However, the World Bank also needs to build its own capacity to take advantage of the potential of converging technologies for accelerating human capital outcomes, while ensuring inclusion and empowerment. In doing so, it will have to address several internal constraints. Although the study noted the widespread take-up of technology across many projects, on the whole this approach still tends to be piecemeal and pilot-focused as opposed to systemic and at scale. To keep up with the rapidly changing technology and innovation environment, the skill mix of staff should be continually upgraded. Stepping up engagement and collaboration with country and global experts should be elevated in priority. In particular, partnerships with leading innovation hubs and a capacity to critically evaluate innovative solutions offer promising ways in which to rapidly test and enable policy measures going forward.

This study proposes three broad sets of actions:

1. Develop partnerships so that the Bank becomes a better-informed practice leader.

    - **Develop partnerships.** Identify and network with external thought leaders and leading practitioners on technology for human capital; organize a virtual advisory panel; engage with development partners to coordinate digital transformation programs in health, education, and social protection; pursue engagements with private sector partners and multilateral and multi-stakeholder forums to both explore opportunities for collaboration and shape a normative consensus.

    - **Build sustained client engagement on technology.** Conduct joint scenario exercises with clients; codevelop sandboxes, pilots, and programs with clients; and offer ongoing analytics and just-in-time technology assessments.

    - **Advance the global HD agenda on COVID-19 response and recovery.** Address broader resilience through technology (such as adaptive social protection and local innovation capabilities) to include preparing for future shocks such as new pandemics, climate change, food security, and other major disruptions.

2. Fast-track a shared understanding of a technology-enabled human capital program.

    - **Internalize the findings of this study** by soliciting feedback on the technology landscape and identify areas of broader engagement. Incorporate the nine technology action areas into the policy dialogue and operational projects as part of the South Asia Human Capital Plan in order to improve service delivery, improve resilience and adaptability, and promote inclusion and empowerment. Core elements of the HD agenda should be, for example, setting standards for equitable digital access in health, education, and social protection and jobs; promoting trust in the use of technology; encouraging local digital content; and establishing a technology and data governance framework. Undertake a "one HD approach" to technology and move beyond pilot-focused initiatives to systemwide engagements at the country level.

3. Develop HD's service offerings and raise the capacity for technology design, advice, and delivery.

    - **Develop new service offerings,** including assessments of the technology landscape for human capital and of policy and institutional aspects (such as part of a human capital public expenditure and institutional review); preparatory work on key policy actions and other operational entry points; virtual advisory services focused on technology transformation (such as inclusive AI in HD and data policy for HD); and operationalization of the "no harm" approach (such as through normative scenario exercises, development of standards in targeting, verification of algorithms, and application of ethical AI design principles to datasets).

- **Initiate a dynamic joint learning approach with clients and staff,** building awareness, demand, and capabilities for new skill sets; codesign solutions to real-life problems; and ensure frequent client feedback to stay abreast of changing frontline demand.

- **Articulate a technology-enabled agenda for human capital development,** which would include identifying and developing priority areas for collaboration and reform and establishing a monitoring and evaluation framework and methodology on technology for human capital.

- **Develop systems for feedback from lessons and insights** gained from implementation and learning from stakeholders to improve and develop future action areas as appropriate because converging technologies and their applications, impacts, and governance continue to evolve.

Note

1. The three subheadings are consistent with the three dimensions of human capital development outlined in the World Bank's "South Asia Human Capital Business Plan": (1) improving service delivery through increased and smarter public investment in health, education, and other sectors that contribute to human capital; (2) building resilience and adaptability to future shocks and risks; and (3) including and empowering the vulnerable sections of the population, especially girls and women (World Bank 2020).

Reference

World Bank. 2020. "South Asia Human Capital Business Plan." World Bank, Washington, DC.

Technology Metatrends

TABLE A.1 **Metatrend 1: Technologies for Building and Protecting Human Capital**

Impacts of disruptions caused by COVID-19	Technology trends
1. The pandemic is causing deep disruptions in daily routines and the social organization of family life, school, and work. Beyond the immediate health emergency, a strong negative impact is the side effect of lower economic growth, resulting in higher poverty rates and loss of life from malnutrition and other diseases. 2. As countries across the region maintain extended lockdowns, health systems and social assistance programs are strained to save lives and protect livelihoods. 3. With millions of children out of school, there is rising concern about long-term learning losses and permanent school dropouts, especially among girls. 4. COVID-19 lockdowns are resulting in a temporary halt in ongoing nutrition, welfare, and treatment programs, leaving many people without access to health care and causing a long-term increase in mortality. 5. Food security remains a source of concern over the medium term and is testing the resilience of supply chains.	1. Expansion and upgrading of mobile access. 2. In response to school closures, many education systems scaling up distance learning programs, using radios, television, and SMS (short message service) channels for instructions. Choice of education technology varies according to location, connectivity, and affordability. 3. Renewed focus on expansion of digital platforms for delivery of health and education services. The transition will be affected by requirements for system integration, development of new teaching content, questions about learning impact, and review of national testing standards. 4. In parallel, new offerings vying for user adoption, facilitated by digital micropayments for health, education, insurance, and welfare services. 5. Increasing differentiation among providers, with pioneers at the frontier offering personalized and immersive learning at a premium. Travel and visa restrictions causing a sharp reduction in study-abroad programs. 6. Rollout of training programs for teachers and health care professionals to help with the adoption and use of new technologies. Programs will also capture data, track performance, check eligibility, and verify attendance.

(Table continues on next page)

TABLE A.1 **Metatrend 1: Technologies for Building and Protecting Human Capital** *(continued)*

Impacts of disruptions caused by COVID-19	Technology trends
	7. Telemedicine expanding access to primary care, particularly for remote communities, easing the shortage of qualified health professionals. Medical equipment shortages are prompting local responses, such as 3-D printing of personal protective equipment. Over the medium term, telemedicine offers opportunities for upgraded medical care. Machine learning and artificial intelligence (AI) will improve disease diagnoses, leading to better health outcomes and improvements in human well-being.
	8. More granular solutions (such as solar, microhydro, and battery storage) offering reliable, cost-effective energy sources, especially for households. Better access to energy remains a key building block for a desirable quality of life and well-being by enabling improvements in education, health, food production, clean water, sanitation, air quality, resilience to climate change, security, and safety.

Metatrend 1: Potential Implications for Development

Positive	Negative
1. The COVID-19 crisis has demonstrated that humans are resilient and adaptable, provided that suitable technologies are available.	1. Lack of anticipation of inclusion, democratization, and community involvement may deepen existing inequalities between privileged elites and digitally underprivileged groups, resulting in more fragmentation across society.
2. Rapid exposure to technology applications will increase digital adoption and offer alternatives to overcrowded school and health facilities.	2. Lack of data about online learning makes it difficult to assess outcomes and identify new ways to combat shortfalls.
3. New production possibilities for education, knowledge, health care, and social assistance may expand access to opportunities.	3. Lack of training and technical support may lead teachers to feel overwhelmed and oppose the use of new teaching formats.
4. New entrants promise to expand the affordability and accessibility of primary education and health care.	4. The additional burden borne by mothers and working parents and the cost of online learning may aggravate psycho-emotional stress on families and lead to a long-term decline in nurturing and learning outcomes for children.
5. Learning communities can provide mentorship and human-centered approaches to converging technologies (such as 3-D printing, interactive games, and local content development).	5. Digital tracing applications and data mashups raise complex trade-offs among protection, social control and surveillance, bias, cyber risks, and manipulation.
6. Growing awareness among young people of the need to develop adaptive skills and interdisciplinary literacy—cyber-info-security, bio-preparedness, and resilience.	6. Without investments in human capital infrastructure, such as water and sanitation, energy and environment, mobile connectivity, and sustainable food production, South Asia could face steeper declines in living standards, divergence in opportunities, and greater inequality. Are available no-tech and low-tech solutions being crowded out by digital investments?
7. Returning migrants and tech entrepreneurs in the diaspora help accelerate technology-based initiatives.	

Source: World Bank study team.

TABLE A.2 Metatrend 2: Data-Driven and Hybrid Human-Machine Technologies for Productive Activities

1. Global lockdown has interrupted global supply chains, accelerating pressures on wages, reshoring, and automation.
2. New production technologies will reconfigure digital supply chains and the demand for labor-intensive exports from developing countries. With few large-scale export sectors in the South Asia region, industrial activities may increasingly shift to producing for local consumption.
3. Digital technologies and innovations will disrupt production processes in nearly every sector: agriculture (precision farming), transport (self-driving cars), manufacturing (robotics, 3-D printing), retail (e-commerce), finance (e-payments, AI-driven trading), media (social networks), health (AI diagnostics, telemedicine, drug discovery), education (online learning, virtual classrooms), and public administration (e-governance).
4. The role of data will increase, both as a resource and as a source of power. Driven by global competition, the South Asia region will see increasing use of AI, robotics, and additive manufacturing, services, and knowledge work. Data-enabled machines and processes will become more deeply integrated with the knowledge economy and replace tasks once performed by humans. A new blend of human-machine interaction may improve productivity but displace many low-skilled jobs. In a world in which humans and machines compete for cognitive performance and societal relevance, new definitions of agency and self-determination may be needed.
5. Employment prospects in the information and communications technology, outsourcing, and freelance sectors remain strong in parts of South Asia, facilitated by new work styles and ubiquitous digital services.
6. Digital entrepreneurs, including in diaspora networks, are mobilizing funding and technical solutions.
7. The new and different jobs emerging are rooted in human abilities (such as judgment and creativity), interpersonal skills, and compassion.
8. Modern management practices can leverage disruptive technologies to help firms improve performance and enable innovations. Competitive pressures will push firms into adult education, expanding new forms of on-the-job training.
9. Workers in both formal and informal jobs will need to reskill and upskill frequently to keep up with the evolution of technology.
10. AI-enabled digital platforms can match employers and job seekers with high accuracy and quickly bring on board "gig" workers for temporary assignments.
11. Social distancing will have a lasting impact on the physical distribution of work (such as the future design of urban spaces, slum upgrading, and expansion of secondary cities).
12. Digital employment records will be integrated and linked with e-commerce, digital finance, mobility, and health and education services.
13. Second-order effects will emerge. High levels of un(der)employment, particularly among migrant workers in urban areas, are putting pressure on a fragmented social protection system. Demographic pressures, especially a growing youth bulge, may force governments to launch a new generation of employment programs.
14. Uncertainty over employment will intensify pressures for portable minimum insurance benefits.

(Table continues on next page)

TABLE A.2 **Metatrend 2: Data-Driven and Hybrid Human-Machine Technologies for Productive Activities** *(continued)*

Metatrend 2: Potential Implications for Development

Positive	Negative
1. Increase in entrepreneurship. 2. Digitalization, upskilling, and increased human-machine interaction may accelerate innovation (see metatrend 3). 3. A young labor force enables a faster adoption of digital technologies and transition to new production processes. 4. Growing awareness about societies' fragility may have a long-term effect on values and consumer preferences (such as a switch to green energy, mobility solutions, and local food production). 5. Digitalization of the economy may accelerate the transition from predominantly informal activities, allowing people to access markets for services and goods and to participate in new digital activities in the gig economy. 6. New opportunities emerge for home-based work, including for women (although this may reinforce socioeconomic exclusion).	1. Risk of growing economic divergence and rising inequality at the level of nations, firms, and individuals may give rise to economic nationalism and societal polarization. 2. Big tech companies solidify their dominant monopoly positions, which may slow down (local) innovation and intensify a winner-take-all dynamic. 3. The gig economy may increase the economic fragility of workers and impose additional social stress on families. 4. The loss of personal data may not enter most people's awareness, raising fears of a permanent loss of data privacy. 5. Responsible oversight and meaningful accountability in complex technological supply chains will fragment. 6. As automation continues to displace human labor, digital have-nots will find it harder to adapt. Unless new jobs are created in large numbers, growing unemployment and unrest may erode social cohesion. 7. Global trade continues to shrink. The traditional prescription for development through economic growth could come to a halt, limiting the shift of production and jobs to emerging economies and reducing the volume of migration and remittance flows for the region.

Source: World Bank study team.
Note: AI = artificial intelligence.

TABLE A.3 **Metatrend 3: Complex and Dynamic Innovation Ecosystems**

1. Conventional R&D approaches and metrics remain out of reach for most developing countries. 2. The alternative—fostering innovation ecosystems—is viewed increasingly as offering access to diverse stakeholders, expertise networks, funding, and global knowledge as part of a long-term engagement. 3. The world over, governments and firms alike are grappling with how to connect with emerging innovation systems to unlock future drivers of productivity, employment, and competitiveness. Along the way, new forms of collaboration, skill deployment, incentives, organizational capabilities, regulatory approaches, and policies are being tested.	4. The value of a tech-enabled civic culture that relies on bottom-up information sharing, public-private partnerships, "hacktivism" and grand challenges for quick solution testing, and participatory collective action is attracting interest from key stakeholder groups seeking to emulate these approaches. 5. Specialized knowledge institutions, especially in scientific and innovation communities, are being sought out for expert advice in anticipating, preparing for, and responding effectively to crises. 6. South Asian countries are seeking to build their domestic capabilities to participate in the global knowledge system, take advantage of opportunities offered by available technologies, adapt them to relevant domestic needs, and offset some of the risks.

(Table continues on next page)

TABLE A.3 **Metatrend 3: Complex and Dynamic Innovation Ecosystems** *(continued)*

7. Diaspora communities continue to provide local innovators with critical know-how, mentoring, funding, and networks, but government regulations and overreach often stand in the way and fail to create an enabling environment for innovative business models.
8. There is growing awareness that social capital is an important complement to tapping into indigenous knowledge, scaling up grassroots innovation, and developing technologies to address specific local challenges.
9. Participation through local networks and indigenous knowledge communities provides solidarity, resilience, and targeted support at direct beneficiary levels.
10. Product innovations being developed in so-called low resource settings are finding opportunities for reverse innovation.
11. Many countries are expanding R&D and diagnostic and sourcing capabilities to build up national stockpiles of critical supplies in an effort to reduce dependencies and achieve a higher degree of self-sufficiency in, for example, vaccines and medical equipment.
12. New forms of scientific research collaboration are emerging. For example, computer modeling and big data are increasingly being used for drug discovery, which, together with advances in synthetic biology, offer prospects of more affordable new drugs and therapies.
13. Digital technologies—such as smart phones, AI, Internet of Things connectivity, digitalization of information, additive manufacturing, virtual reality and augmented reality, machine learning, blockchain, robotics, quantum computing, and synthetic biology—will accelerate granular innovation processes.
14. Innovations triggered in response to the COVID-19 pandemic have been shown to diffuse rapidly, resulting in substantial performance and productivity improvements.
15. Transparent communication and data-driven decision-making are competing with targeted misinformation campaigns to influence public perceptions, build public support, and encourage actors to contribute to solutions.
16. The emerging lessons from how countries have prepared for and managed disasters are receiving global attention as governments seek to highlight and brand their expertise and prepare for the next crisis.

Metatrend 3: Potential Implications for Development

Positive	Negative
1. The pandemic response is viewed as a demonstration of government leadership and effectiveness. Trust matters! In some cases, effective state responses are increasing trust in government and technocratic expertise.	1. The pandemic revealed innovation gaps in delivery systems for health, education, and social assistance, which hampered effective early response strategies and protection.
2. Civil society groups are mobilizing responses on the front lines of the crisis, indicating innovation capacity and democratic vitality at the local level.	2. The "old" system is not resilient and is inadequate to respond to crisis situations. Part of this reflects the increased inequity in income, wealth, and opportunities and lack of health care and social security benefits for all.
3. Digital technologies can be a powerful mechanism in accelerating learning, overcoming social inequalities, and expanding access.	3. Institutional readiness matters! Unfinished federalism reforms, underfunding of public programs, and confusion over deploying available delivery systems across government agencies have revealed fundamental disconnects in countries' ecosystems.
4. More granular innovation will lead to faster diffusion cycles, lower investment risks, more opportunities to escape lock-in, higher job creation potential, and larger social returns.	4. Technologies are not neutral! Digital technologies themselves are characterized by large disparities in access to, usage of, and the skills relevant to innovations.

(Table continues on next page)

TABLE A.3 **Metatrend 3: Complex and Dynamic Innovation Ecosystems** *(continued)*

Metatrend 3: Potential Implications for Development

Positive	Negative
5. The mobilization of whole-of-government and whole-of-society responses will require long-term investments in social and political capital, including through community participation, an agile response capacity, and accountability mechanisms. 6. Governments are taking a renewed interest in steering innovation to tackle societal challenges.	5. The broader development benefits from using digital innovations have fallen short of being inclusive and are unevenly distributed. Most of the population lacks the capital needed for innovation. 6. Converging technologies raise many challenges in terms of increasing inequality, loss of privacy, loss of agency, and loss of freedom (see metatrend 4).

Source: World Bank study team.
Note: AI = artificial intelligence; R&D = research and development.

TABLE A.4 **Metatrend 4: Governance of Dual-Use Technologies**

1. COVID-19 is leading to a rapid expansion of executive powers, including through the deployment of data-driven technologies, with potential implications for democratic spaces (such as freedom of movement, states of emergency, and postponement of elections).
2. The development of integrated digital and biometric identification (ID) systems will accelerate the expansion of civil registries to access public services and activities (such as social welfare, health coverage, education, mobile phone, digital finance, internet access, and voting).
3. The capacity for using citizens' personal data for bio-surveillance and behavioral monitoring (such as contact tracing and social media use) will continue to expand. The combination of digital surveillance and behavioral manipulation may lead to new demands for data privacy protection.
4. Nonstate actors who benefit from control technology, data pools, and the accumulation of new digital wealth will exert growing influence over the daily lives of citizens, the cohesion of societies, and the economic prospects of countries.
5. As all activities in a society become digital, the accumulation, control, and use of data will remain a contentious issue at the national and international levels between state and nonstate actors.
6. Universal connectivity of devices and convergence of technologies will expose new cybersecurity risks and vulnerabilities to cybercrime. The potential of data wars becomes a primary concern for national and international security.

7. As technology leaders race ahead, the risk of deepening divides and power shifts between groups with technological capabilities and those without is growing.
8. The convergence of AI and biotech holds the distinct prospect for changing how genomes of humans and other species are computed, designed, and programmed.
9. The concept of governance continues to evolve. With AI deployment (and its built-in biases), regulation can take the form of codes, ethical commitments, or corporate principles. Reliance on algorithms for decision-making without appropriate safeguards (such as human-centric design and oversight) will raise new questions of human accountability.
10. As technologies continue to evolve and generate new forms of knowledge and decisions, human capabilities do not necessarily stretch as far, with unknown consequences for the adoption, use, and outcomes of technologies. Unless human-centric design and oversight are ensured, human accountability for machine-based decisions will be in doubt.
11. The collaboration (and, in some cases, mutual dependence) between big government and big tech may put individual rights at risk and expose citizens to manipulation and targeted misinformation. The growing presence of social media platforms and the ability of national ID systems to track every citizen concentrate data and power in the hands of a small elite.

(Table continues on next page)

TABLE A.4 **Metatrend 4: Governance of Dual-Use Technologies** *(continued)*

12. Regulation and oversight of technologies will become the next frontier, pitting secrecy and control against calls for multistakeholder engagement to discuss societal norms and a new digital social contract. 13. Responsible governance of converging technologies within sociotechnical systems may create new opportunities for addressing poverty (such as the AI for Social Good movement).	14. There is a growing reliance on algorithmic decision-making, raising the risk that biased data and faulty modeling assumptions may deepen inequality and institute systemic discrimination. 15. The open-source software movement is gaining momentum for utilizing decentralized technologies to innovate and operate without official permission.

Metatrend 4: Potential Implications for Development

Positive	Negative
1. Prospects are enhanced for collaboration among public, private, and civil society actors to shore up normative principles so that the new digital ID systems provide access to social protection while ensuring data security, privacy, and inclusion for all, including the most vulnerable groups. 2. Public and private actors could collaborate on inclusion and sustainability agendas and invest in infrastructure programs that empower households. 3. Democratization of AI, additive manufacturing, and biotechnologies are empowering local communities to innovate with their own data and designs in alignment with their needs and ethical norms (such as citizen science and FabLabs). 4. New opportunities are arising for identifying biases in datasets and using social media and grassroots coordination to lobby for reforms. 5. The pandemic may spur innovations in electoral and voting processes that ensure greater preparedness for future shocks. 6. New calls are being heard for digital cooperation and oversight mechanisms.	1. In some countries, political leaders are taking advantage of the COVID-19 crisis to weaken checks and balances, erode mechanisms of accountability, postpone elections, and weaken citizens' fundamental rights. Under the guise of fighting misinformation, control over free expression and the media is being tightened. 2. Converging technology platforms (such as AI, facial recognition, biometrics, and personal and consumption data) are creating new forms of social and behavioral control and eroding privacy and agency. In the absence of a data privacy protection, accountability, and redress mechanism, individuals and groups can become targets of discrimination, exclusion, and political repression. 3. Difficulty is encountered in tracing biases in complex, multidimensional datasets. 4. Data are captured for non-control reasons without consent. An example is medical data. 5. New forms of cyberattacks corrupt the integrity of public and personal data, erode public trust, and hamper government's core functions. Digital manipulation (such as hate speech and misinformation) undermines trust in government's ability to protect its citizenry.

Source: World Bank study team.
Note: AI = artificial intelligence.